Industrial Electric Motors: Installation, Running, Advanced Maintenance and Reliability

Mohammed Hamed Ahmed Soliman

Published by personal-lean.org, 2024.

While every precaution has been taken in the preparation of this book, the publisher assumes no responsibility for errors or omissions, or for damages resulting from the use of the information contained herein.

INDUSTRIAL ELECTRIC MOTORS: INSTALLATION, RUNNING, ADVANCED MAINTENANCE AND RELIABILITY

First edition. July 8, 2024.

Copyright © 2024 Mohammed Hamed Ahmed Soliman. ISBN: 979-8332783326

Written by Mohammed Hamed Ahmed Soliman.

Industrial Electric Motors
Installation, Running, Advanced Maintenance and Reliability

Industrial Electric Motors: By Mohammed H.A. Soliman

Contents

Checking at First Time Upon Reception 6
Installation 10
Drive Coupling 20
Running 27
Starting 32
Maintenance 36
    Routine Periodical Maintenance 37
    Troubleshooting Common Motor Problems 44
    Electric Vibration 45
    Mechanical Vibration 47
    Continue Troubleshooting Causes of Motor Failures 52
    Problems of Poor Lubrication & Bearing Grease 55
Bearings Maintenance 56
    Bearing Diagnosis 63
    Choosing the Right Grease 67
    Assembling and disassembling 69
    Sealed ball bearing 73
Electric Motor Reliability 74
    Criticality of Electric Motors 75
    What is risk management? 78

- Asset Criticality Assessment 82
- **Failure Mode Effect Analysis FMEA as a Great Tool for Minimizing Risks of Failure and improving Reliability** 88
  - Failure Mode 91
  - Example – Failure Modes for Common Equipment 92
  - Overview on the FMEA Process and How to Conduct It 105
  - Improving the Reliability of a Pump Station System 128
- **Electric Motor Predictive Maintenance** 149
  - Infrared Thermography 149
  - Ultrasound Analysis 156
    - Ultrasonic Condition-Based Lubrication 158
  - Vibration Analysis 168
    - On Site Tools Used for Measurements & Analysis of the Mechanical Vibration 181
    - Measurement Techniques (points of measuring) 185
    - Vibration Standards 188
    - Quick Example – Centrifugal Fan 189
    - Example.2 Discuss the Following Vibration Analysis Data Report 191

Methodology of Measuring "Collecting Data" 195
The Different Types of Vibration Sensors 197
Vibration Sensors Connection Types 203
Vibration Analysis-Signal Processing 213
Time Waveform 214
Enveloping and Demodulation Spectrum 216
Phase Reading 217
Frequency Spectrum 217
What is FFT? 220
Spectrum Analysis & Faults Diagnosis 221
What are the Faults that Spectrum Analysis Can Tell Us About? 223
Faults to be Detected by Spectrum Analysis .. 225
Predefined Spectrum Analysis Bands 226
Construction of Spectrum Analysis 227
Detect electric Motor Faults by Vibration Spectrum Analysis 230
   1. Detect the Nature of Bearing Failure 233
   2. Detection of Different Electric Problems 238
      A. Rotor Defects 240
      B. Eccentric Rotor 242
      C. Stator Defects 243

- D. Phasing Problem (loose connector) 244
- E. Synchronous Motors (Loose stator coils) 245

3. Using Vibration Spectrum Analysis to Detect Machine Looseness 247

- I. Internal Assembly Looseness 247
- II. Looseness between Machine to Base Plate 249
- III. Structure Looseness 250

Literature References 254

Industrial Electric Motors: By Mohammed H.A. Soliman

Checking at First Time Upon Reception

These rules are for squirrel-cage motors with high or low voltage. It doesn't matter if they are foot-mounted or flange-mounted, as long as they are drip-proof or totally enclosed fan cooling type. For motors made differently or for specific uses, please check the booklets that come with them for details.

Check Upon Reception

- Please make sure all the items listed on the receipt are correct.
- Please read the test results closely.
- Look to see if the motors are broken or dirty, and fix them if needed. Make sure everything is there and nothing extra is in it.
- Start the motor and check if it spins in the correct direction.
- Show if something is suggested. The way to go is always shown with a sign

- showing an arrow that's connected to the motor.
- Carefully read the ratings on the main nameplate and other plates.
- Turn the shaft by hand and check that it turns properly. To stop from happening (To prevent rotors from moving and protect the roller bearings, the 2-pole motor shafts and motors with roller bearings are stuck while moving goods. Please open them before turning.)
- If you need something specific, like a certain color, please make sure the paint and accessories are okay.

If you find any issues, please get in touch with the manufacturer. Get the address of the company or the closest service station to you and provide them with the following information. Information: The kind, number of poles, power output, voltage and frequency displayed on the label. The nameplate shows the test number and manufacture number. They can be found there (labeled on the nameplate or the end of the shaft).

The one way of locking the rotor.

The other way of locking the rotor.

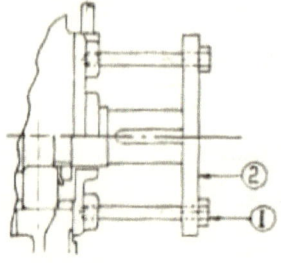
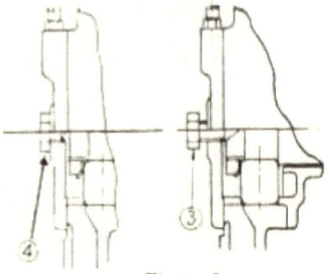

Figure 1.
Please unlock the bolts (1) and take away the shaft-end plate (2) before running the motor.

Figure 2
Replace bolt (3), shown in the drawing to the right, with bolt (4), shown in the drawing to the left, before running.

If you make machines, please read this:

If the motor rotors you got are stuck, the purpose stated above is to lock the rotors again. You need to do this by the methods explained above or by the methods shown below:

Industrial Electric Motors: By Mohammed H.A. Soliman

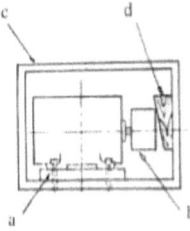

a. The foot of the motor is locked to the bottom of wooden case.
b. Belt pulley or other transmission device.
c. Wooden case.
d. Wedge of hard wood or other proper material.

Figure 3

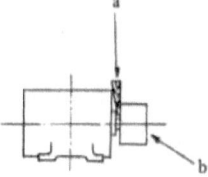

a. Wedge.
b. Belt pulley or other transmission device.

In case, you choose this way for locking the rotor, you must make sure that the belt pulley or other transmission device will not get loose during transport.

Installation

A. Installation Environment

A good environment plays a big part in being good operation and functionality. To make sure the motors work well, you need to think carefully about it when you write down what you want to buy. To make sure things are safe to use, here are some things for you to know:

1. Temperature of the surrounding air or place

1.1. Standard motors work best when the temperature around them is between -15°C and 40°C.

1.2. If the temperature around the motor is too hot, we need to use cooling or heat-insulating methods. We can also reduce the load on the motor to prevent it from getting too hot.

1.3. Conversely, if the temperature is very cold, we will need to use heating to make it warmer.

1.4. If the motor is used in a place that is either very hot or very cold, the insulation, wires, oil, bearings, and other parts of the motor may need to be checked and possibly changed.

2. Good circulation of air (good ventilation)

2.1. If the air doesn't flow well to the motor, it will get too hot.

2.2. Please make sure to keep items at least 20cm away from the air intake ports.

3. A place with not enough fresh air circulating (bad ventilation).

If the motor is put in a space with not enough air, we need to make it better so the motor doesn't get too hot.

4. If the motor is put outside where it's wet or there's water, we need to protect it from getting too hot.

5. Dust: In a very dirty place, there may be problems that come up, so it's a good idea to clean it regularly.

5.1. Open type

Means a format or style that is available for anyone to use or access.

A lot of dust on the coils and pipes in the center will make the coils get too hot. Also, dust and moisture trapped by it might lead to the insulation not working properly. If there's too much dust on the rotors and it's not spread out evenly, it can cause the rotors to be unbalanced and make the car vibrate. If dust gets into the bearings, they can get broken.

5.2. Sealed type

A lot of dust on the fins of the frame will make it harder for the heat to go away. If there is too much dust on the fan or transmission device and it's not spread out evenly, it might shake and make a noise.

6. Harmful gases and hot water vapor.

If there are harmful gases or steam in the air, you should use motors that are explosion-proof or resistant to corrosion. If you can't do that, you need to take safety precautions. We need to be very careful when choosing a motor, especially if it will be used in places with flammable gases, dust, or steam. It's important to make sure the motors you choose meet safety standards and are built to prevent explosions.

7. Accessible Site

Places where motors are installed should be easy to get to and open so that the motors can be easily brought there and installed. Also, doing tasks like checking, cleaning, and fixing (especially adding grease) will not be limited.

8. Base that cannot be affected by shaking or movement

8.1. Engines need to be put on a strong and sturdy base or floor that won't shake from the surroundings.

Industrial Electric Motors: By Mohammed H.A. Soliman

8.2. Strong shaking from the surroundings can cause harm to the motors installed there.

8.2.1. Roller bearings can get depressed when the motors are not running for a while.

8.2.2. The windings can get broken.

8.2.3. The covering on the wires and connections might get broken.

8.2.4. The ground where motors are placed must be solid and firm.

<u>9. Steady; or else, the shaking may get worse and worse, especially with powerful machines like crushers and compressors. Big shaking while the motor is on can cause these problems:</u>

9.1. The covering on the coils may be broken.

9.2. Bearings may not last very long.

9.3. Some parts might become loose or move out of place.

9.4. The cooling fan or other rotor parts might stop working because they are worn out.

10. Power supply

10.1. The voltage coming in should stay steady and not drop too much when using power.

10.2. If you need to use a different voltage and frequency to power the motor, look at sections 4. 2C and 52C for help. If it's not really needed, please don't do it because the motors can get too hot and not work well.

11. Height of the Installation

If the sites are higher than 1,000 meters above sea level, the temperature of the motors operated there will be 5 to 10 degrees Celsius higher.

B. Foundation and Installation

We won't talk about how to lay the foundation here. We want to tell you about how to install the motors. Read the following descriptions carefully.

1. The motors need to be put into the concrete or foundation so they can run smoothly. Do not curve or turn the bottom. Put the Packers under the motor's feet and the bearing stands to support their weight. We need to put packers on each side of the bolts at the base. The packers should be 300 to 500mm apart to evenly support the weight. It is best to use shrink-resistant mortar and grout. The bolts at the base cannot be tightened until the mortar around them is completely hard. A lot of grouts should be put under the base to make the foundation stronger.

2. Foundation and base for 2-pole motors

2.1. Since the synchronous speed of 2-pole motors reaches as high as 3,000 or 3,600rpm, if their foundation is not well designed and constructed, resonance may be resulted in. Be very careful, please.

2.2. Natural frequency and resonance

National frequencies of 2-pole motors are listed below:

	FIRST	SECOND
In 50Hz area	3,000cpm	6,000cpm
In 60Hz area	3,600cpm	7,200cpm

If the natural frequency of any other object in the installation equals or approximates any one of the values listed above, resonance will be resulted in.

If natural frequency (fn) is considered in one free degree, it may be expressed as:

$fn = 1/2\pi \times \{(g \times K)/W\} \wedge (1/2)$

where g=gravity acceleration

W=weight of coupling machine

K=elastic coefficient of the system

The values of W and K should be carefully chosen that the

value of fn will not equal or approximate any values listed above. If resonance is noted or suspected when the motor is running, please measure the natural frequency and change it.

3. Please make sure to inspect the items after the foundation is done and the motor base is installed.

3.1. Please check and confirm that the foundation is as strong as mentioned in 2.1H and 21I

3.2. Ensure the foundation is strong enough to reduce the vibration of the motor and its connected machine.

3.3. Inspect and verify that the building is sturdy and will not sink or shift due to changes in the ground and foundation.

3.4. Please check and make sure the concrete buildings have finished getting smaller or changing shape.

3.5. Ensure that the foundation and base are flat and even. Machines that work together should also be flat and even in relation to each other, with a difference of no more than 0. 2mm

3.6. Ensure there is room for base bolts, cable wire, thermometers, space heaters, and pipes for conductors. Also, locate where these items are located.

Drive Coupling

Centering

When the motor is connected directly to the machine it drives, it is important to make sure they are properly aligned. First, use a level to make sure the bottom of the motor's foot is all flat and even. Next, use bolts to attach the foot to the base for a test installation. Next, place the dial indicator on the side coupling device and slowly turn the motor shaft to get the exact measurement of A as shown in the figure below. Next, check the angle between the motor shaft and the machine it's connected to. Use a tool to measure the gap, and if needed, add or remove spacers to make sure everything is aligned properly. Try to get the gap as small as possible.

Industrial Electric Motors: By Mohammed H.A. Soliman

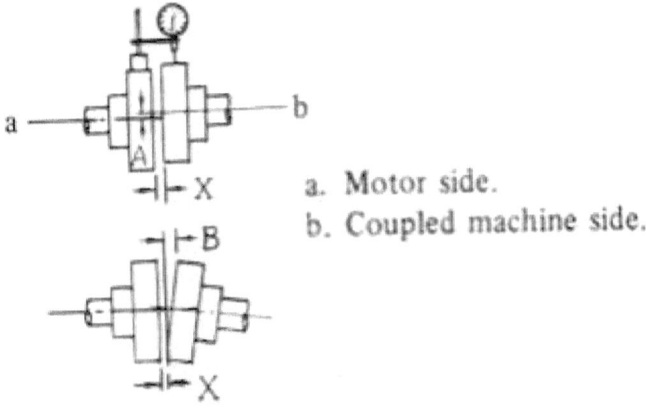

a. Motor side.
b. Coupled machine side.

The allowance of A, B and X are listed below:

	Rigid Coupling Device	Flexible Coupling Device
A	0.03mm	0.05mm
B	0.03mm	0.04mm
X	0	The value designated by manufacturers

However, gear couplings or specially designed flexible coupling devices may have more flexibility than the ones listed above. Please ask the companies that make the products for more information.

Belt Transmission

V-belts are used to transmit power from motors.

In our discussion, we will use these belts as an example because they have average abilities.

The belt pulley on the motor is 8 times larger than the one on the machine it's connected to.

The V-belt should not go faster than 22 to 23 meters per second. If the belt goes faster than 25 meters per second, it will start to slip and vibrate more. This will make the belt wear out faster.

If the pulley is too small, the force on the shaft will be more. If this force goes too high, the shaft will break. if you need to use a small, wide pulley for a belt to turn something.

When it comes to using belts to transfer power, these steps will be followed:

1. To make sure the belts work well together, it's best to get them all from the same company. If that's not possible, make sure the size difference between the belts is less than 0. 2% Remember to choose the right belt. If any belts on the same machine are broken or old, you should put all new belts on, not just the ones that are broken or old. Next, we need to make sure that the pulleys are properly lined up. Keep the motor shaft in line with the shaft of the machine it's connected to. Or, if using belts or pulleys, make sure the shafts are at a right angle to each other.
2. Ensure the belts are not too tight. If they are too tight, the shaft may get damaged or broken. Alternatively, if the tension is too low. The belts might come off, the power won't work well, and the belts and pulleys will wear out faster. You can easily figure out how tight the belts are by following these

steps: Hang a weight P in the middle of the two belt pulleys and measure h1 and h2 as shown in figure below.

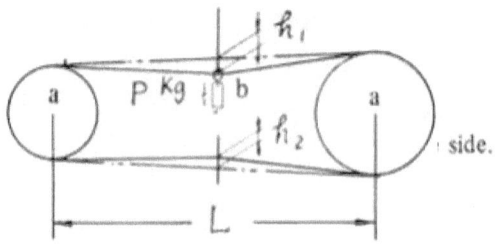

a. Belt pulley.
b. Spring balance

The values h1 and h2 show how much the deflection changes when the weight is hung up. The rough values of. The equation e=2×L/(h1+h2) is displayed in the table below:

Shape of V-belt	P (kg)	e	Transmission condition	
			Speed of belts	Output
C	5	35	16m/s	30Kw
D	10	35	25m/s	150Kw
E	10	45	25m/s	150Kw

You can turn the bolts at the bottom to make it tight. The values of e go up and down with the speeds of belts and the amount of output. We need to change the values to keep things working right. Another way to measure the tension is to measure how much the belt stretches. When the belt is stretched by 0.5%, it means it has been put on with the right amount of initial tension. After the motor has been running for 10 hours, check the belt tension. After the motor has been running for a month, the belt might be stretched out and need to be adjusted.

3. Changing V-belts

If the V-belt is worn out, it needs to be changed. If not, the transmission won't work properly. Furthermore, because the amount of friction is reduced, the shafts will be stronger and more resistant to bending. That will harm the bearings. Usually, when the belts stretch between 1.5 - 25%, they need to be changed.

4. Special belts
Are belts that are different or unique in some way. Right now, a lot of nylon and steel belts are being sold. If you use these tight belts, the motor shafts will need to be very strong and flexible.

Industrial Electric Motors: By Mohammed H.A. Soliman

Running

Make sure to check before you start:

Please read these descriptions before you begin.

1. Distribution line means a line that carries electricity to different places like homes, businesses, and other buildings.

2. Please make sure to verify the power supply, magnetic switch, and other safety devices, Star-Delta Starter, reactor, compensator, wires for space heater and thermometers, and the distribution line for other machines.

3. Insulation resistance is how well a material can resist electric current traveling through it.

4. Ensure that the wires are securely connected, the insulation is in good

Industrial Electric Motors: By Mohammed H.A. Soliman

condition, and the terminals are spaced properly.

5. Grounding means connecting to the earth's energy. It can help people feel calm and centered.

6. Please check that the frame or terminal box of the motor is connected to the ground.

7. Insulation Resistance means how well something stops electricity from flowing through it.

- To check the wires connected to the stator and rotor by using their terminal connectors.
- Use a 500V testing device to test stator windings that are under 3KV. Test the stator windings that are above 3KV with a 1000V device. "All the rotor windings are connected to a 500V device."
- The ability of insulation to resist electricity changes with the amount of power being used. Rewrite this text in

simple words: "and voltage." "and how strong the electricity is." Categorizing how well something keeps heat in and how fast the motor spins. However, it also changes depending on the temperature, wetness, dusty conditions, how long it's been used, testing voltage, and how long it's been tested. Because of the insulation resistance, we can't measure the resistance (R). But this rule will be used.

3M ohm for rated voltage above 600V.

1M ohm for rated voltage below 600V.

or, formulae from JEC - 146 may apply;

$R \geq$ (rated voltage) /rated output (KW)+1000(M ohm)

$R \geq$ { (rated voltage + rpm/3) / rated output (KW) +2000 } + 0.5 (M ohm)

- If the insulation resistance gets low, the windings need to be dried using hot air, vacuum, or electrical current. This depends on the situation. If the resistance does not reach the level we want after drying, there may be some problems. Please find and fix the problems. If the job can't be done, please contact Tatung company or the nearest service station.

8. Lubrication

    ✓ Engines get lubricated when they are put together. However, still It may take a while for the things we put together in our factory to start working in your factory. We need to add some more parts before it can start running. When adding more grease, please use the same or similar products. For special motors, the type and amount of grease to use are written on the nameplates.

- ✓ Check to make sure the machine is working well. Make sure to also check the connection, the tightness of the belts, and the fastening devices.
- ✓ The rotors of 2-pole motors or motors with roller bearings have frequently been locked during shipping, so please check that it can be turned easily. Inspect and make sure there are no objects inside or entering the motor during assembly or shipment.

Industrial Electric Motors: By Mohammed H.A. Soliman

Starting

1. Start Load: Usually, motors are turned on without any work or load to test them. First, it needs to be tested to make sure it is working correctly. Then it can be connected to a machine for more testing. Engines are usually started with a light load and then switched to a full load once they reach full speed, unless there's a specific need to do otherwise.
2. Direction of Rotation: When you look from the opposite side of the one that powers the machine, the right way for it to turn is clockwise. After the connectors on the end of the wire U. V When Wires W (or 1, 2, and 3) are connected to connector R, S, and T from the power supply, the motor turns in a counter-clockwise direction. If you switch any two of the three connecting wires, it will make the motor turn in the opposite direction.

Most engines can spin in either direction. But, 2- or 4-pole motors with fast speed or 6-pole motors with big power, can only spin in one direction, either clockwise or counter-clockwise. In this situation, a plate with an arrow showing which way the motor turns will be put on.

3. Supplied Voltage and Current: A. Make sure supplied voltage agrees with or within ±10% of that shown on the nameplate. If the difference between rated voltage of the motor and supplied voltage, the windings may become overheated. B. Make sure supplied 3-phase voltages is in balanced condition among phases. A little bit difference among them, unbalanced currents of considerable values may be resulted in. C. Make sure phase currents are balanced, otherwise windings will be exceedingly overheated and torque cannot be provided. Sometimes,

abnormal noise and severe vibration may be accompanied.

4. Frequency: The difference between the rated and supplied frequencies should be no more than 5% when the voltage is at its rated level. If the voltage and frequency change together, their total absolute value should be less than 10%.

5. Starting: If the first try doesn't work, you can try again. However, as a rule, you can only start the car two times in a row when it is cold. If both attempts fail, wait 30 minutes for the wires to cool down after being heated by the initial current.

6. Starting Time and Noise: If the coupling machine has a big GD^2, it might need more time to start. If you have trouble starting or it takes a long time to start, and you hear loud noises, please call Tatung or go to a service center nearby.

7. Vibration: Find the vibration level using a vibration meter or by feeling it, and then compare it with the data in Section 7 about vibration. Once this is done, allow the motor to run by itself. First, run the machine without anything on it. Then, run it with a full load. If everything is okay at the beginning, run the motor for 3 hours and write down the information every 15 minutes as described in section 5. 2 If everything looks normal, the motor will be kept running and checked every few hours. However, everything is normal, the engine is working fine and can be used.

Maintenance

Routine Periodical Maintenance

Part To Be Checked	Period Of Check	Items To Be Checked		Corrective Steps
Motor Frame	Annually	vibration	Vibration meter, Feeling	Find out the cause of abnormal vibration and corrective steps taken.
	Annually	Noise	Hearing, Stethoscopes rod	Make a survey of the site.
	Annually	Temperature of frame	Feeling or Thermometer	Check power source, load and dust accumulated on fins. Correction be made.
	Annually	Dirt on the frame		
				Check and see

	Frequency	Item	Method	Action
	Annually	Ventilation	Temperature of circulating air, by feeling or thermometer	whether the inlet and outlet of air are blocked. Clean the blinds, air nets and air filters.
Power Source	Annually	Voltage & current	Electrical measuring meters	Check and see whether there is any change of the rated value.
Bearing	Annually	Sound from bearings	Stethoscopes rod	Grease be supplied and removed as prescribed on nameplate. Surface layer and color of the oil. Look up 6.1, 6.2 and 6.3 for information.
	Annually	Temperature of bearings	Thermometer	
	Annually	Vibration of bearings	Feeling or Vibration meter	
	Timely	Discharged grease	Color, hardness and foreign bodies.	
	Annually	Oil leakage		
Inside The Motor	Every 6 Months	Mega ohm of windings	Mugger	Periodic cleaning.
	Annually	Bad odor	Open type machine	
	Annually	Dirt	Dust, foreign bodies and water.	
Others	Annually	Coupling status	Coupling device belts cougars.	Stop the motor for checking.
	Annually	Protective device	Make sure they are in serviceable condition.	
	Annually	Fastening device	Foot-fastening bolts and other fasteners.	

Knock-down examination

For motors running continuously day and night, a knock-down examination should be made every 2 or 3 years. However, for motors designed for special purposes, knock-down examinations should be made in certain time periods prescribed respectively.

Records

Every day records

- ✓ At what time (year, month, day and hour) and in what weather the test is conducted.
- ✓ Voltage. load current. frequency (see 5.2C).
- ✓ Ambient temperature (room temperature).
- ✓ Temperature and noise around bearings.
- ✓ Temperatures in stator windings and on frame surface (totally enclosed type). See 5.2C.
- ✓ Abnormal vibrations and noise.

Records of periodical test and inspection

- ✓ Insulation resistance and the relative humidity
- ✓ Amplitude of vibration
- ✓ The color and contents of grease discharged from bearings.
- ✓ Dirt lodged in and on the motor.
- ✓ Coupling allowance of coupling device; the tension of belts.
- ✓ Fastening bolts for base. foot and other parts.
- ✓ In case of oil lubrication, surface condition and clearness of the oil should be recorded. At the same time, check and make sure there is no leakage.

Some of the values obtained by above-listed tests and inspections are variable; their variable ranges given below:

- ✓ The variation of voltage must be within ±10% of rated voltage. The variation of frequency must be within ±5% of rated value. when rated

voltage is applied. when voltage and frequency vary at the same time, the result of absolute values of the two variations must be within ±10%.
✓ Temperature rise (maximum ambient temperature of 40°C) by TM and RM are listed below:

TM: THERMOMETER METHOD

RM: RESISTANCE METHOD

Part	Insulation Type	A Class		E Class		B Class		F Class		H Class	
		TM	RM	TM	RM	TM	RM	TM	RM	TM	RM
Stator Windings	Types other than TEFC	50	60	65	75	70	80	85	100	105	125
	TEFC	55	60	70	75	75	80	90	100	110	125
Rotor Windings	Types other than TEFC	50	60	65	75	70	80	85	100	105	125
	TEFC	55	60	70	75	75	80	90	100	110	125
Bearing	40°C when test is made at the outer surfaces. 45°C when test is made by inserted thermometer. However, when hot resist grease is used for lubrication, the temperature rise can reach as high as 55°C.										

✓ Vibration Bearings: When their service life is compared with that of windings, the allowable values, when motors are running with load are listed below:
25 - 30μ for 2-pole motors
50 - 60μ for 4-pole motors
70 - 80μ 6-and -more-pole motors

In case the vibration measured exceeds the above listed values, please check and find out the trouble, and corrective measures be adopted soonest.

Industrial Electric Motors: By Mohammed H.A. Soliman

Troubleshooting Common Motor Problems

Industrial Electric Motors: By Mohammed H.A. Soliman

Electric Vibration

Vibrations	Description	Cause	Remedial measures
Vibrations caused by distortion of main magnetic flux	Vibration frequency=2f. Nothing to do with load. Proportional to V^2.	Natural vibration frequency of multiple-nod distorted stator approximates 2f.	Stabilize stator core. Check and see whether stiffness of foundation is sufficient.
Vibration caused by unbalanced main magnetic flux	Vibration frequency=2f or f/pxm (m= 1.2);yielding a noise by 2sf. Stator shakes in certain directions. Severe vibration not proportional to voltage and having nothing to do with load.	Distorted rotor gives non-uniform air gap Perimeterwise-dis-tributed windings are not even Unbalanced rotor gives vibration of great magnitude . Natural vibration frequencies of foundation stand, stator and rotor approximates the frequency of power source .	Repair the rotor to obtain uniform air gaps Adjust windings to obtain balanced flux. Reduce bearing gaps. Install voltage balancing line. Test the stiffness of

				foundation.
Vibration caused by mutual action forces between currents of stator and rotor	Vibration frequency=2f. Large magnitude at starting or with load.	Unbalanced windings (broken wire or unbalanced resistance in secondary circuit .		Balance the windings.
Vibration caused by pulsating torque	Vibration frequency=2f. Vibrating force acting toward the perimeter of rotor.	Unbalanced voltage source. Unbalanced windings.		Adjust the voltage and windings.

Mechanical Vibration

Vibration caused by unbalanced weight	Vibration frequency= n.	Unbalanced residuum. Unevenly accumulated dust. Dried insulation. Eccentric deformation by heat. Worn vanes	Dynamic balancing. Cleaning and repairing. Replace worn vane and cutter.
Vibration caused by bent shaft	Vibration frequency = n.	Bending caused by external force. Deformation by heat.	Straighten or replace bent shaft.
Vibration caused by cylindrically deformed shaft	Vibration frequencies=2n,3n	Elliptic or triangular shaft section.	Repair deformed shaft.
Vibration caused by defective rolling bearings	Vibration frequency uncertain. If vibration is caused by defective rolling surface, frequency=number of balls in the bearing x N (nature number)	Depression caused by external vibration during transport or none running periods. Damaged by over load or worn out through normal use.	Replace the bearing.
	vibration = n or	Shaft not at	

Vibration caused by ill-installed rolling bearing	vibration = n,or uncertain. Larger axial vibration magnitude.	Shaft not at right angle with rolling surface. Distorted bracket.	Repair the shaft. Knockdown for reassembling.
Vibration caused by characteristics of bearings	Frequency uncertain. Larger axial vibration. Nothing to do with rpm.	Non-linear characteristics. Resonant brackets. Excessive gaps.	Pre-compression Reduce gaps Replace bearings Modify fittings, change grease.
Oil-whip	Frequency = n/2. Occurs when speed is 2 times of dangerous speed or above.	Self-excited vibration caused by oil film.	Grease of low viscosity. Reduce the width of bearings. Enlarge bearing gaps. Check the diameter at the neck of shaft.
Vibrations caused by distortion coupling devices of driven machine	Frequency = n. Vibration disappears when uncoupled.	Mal-alignment between shafts. Insufficient straightness of shaft.	Adjust the coupling Re-alignment of coupling device.
Vibration caused by improper installation (resonance with installation system)	Vibration frequency equals n, 2n; f, 2f.	Motor base and foundation are not properly coupled with driven machine. Resonance between vibration system of both sides.	Adjust the installation system. Change natural frequency of the system.

f: frequency s: slip p: pole pairs m: integer n: rpm

Tips:

If the motor shakes a lot while running, turn off the power and check if the shaking is because of the motor or because of the electricity. Then, change the weight on the motor while it's running or let it run without any weight to find out what's making it shake.

Measure and change how often something naturally happens.

Testing how quickly something natural vibrates by hitting it.

A vibration meter attaches to the motor's upper part on one side. Then crash the motor at the opposite side of the meter, like shown in picture 22. The test needs to be done in both directions, up and down, and side to side. The test should show natural frequencies that are in the specified ranges.

In areas with a 50Hz frequency, the vibration levels are below 2000cpm, between 3600-4800cpm, and above 7400cpm.

In a place with 60Hz frequency, the levels are below 3000 counts per minute, between 4400 and 6000 counts per minute, and above 8000 counts per minute.

If the natural frequency is too low or too high, we can change the base cotter or add more grout to adjust it. This will help to bring the natural frequency to the right level. We suggest using grout instead of concrete because concrete shrinks.

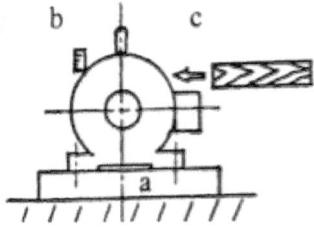

a. Base
b. Synchronous Vibration Meter.
c. Strike with wooden slab.

Industrial Electric Motors: By Mohammed H.A. Soliman

Continue Troubleshooting Causes of Motor Failures

Industrial Electric Motors: By Mohammed H.A. Soliman

Kinds of Breakdown	Symptoms	Possible Causes	Remedies
Fail to start without load	Motionless and soundless	Power-off	Consult power company
		Switch-off	Switch-on
		No fuse	Install fuse
		Broken wiring	Check wiring and repair
		Broken lead	Check wiring and repair
		Broken windings	Check windings and repair
	Fuse blowing. (Automatic switch trips off, slow start with electromagnetic noise)	Short circuit of circuit switches	Check circuit switches and replace
		Incorrect wiring	Check wiring according to nameplate
		Poor contact at terminal	Lock tightly
		Windings grounded	Factory repair
		Broken windings	Factory repair
		Poor contact of circuit switches	Check and repair
		Broken wiring	Check and repair
		Poor contact of starting switches	Check and repair
		Short circuit of starting switches	Check and repair
		Incorrect connections of starting switches	Connect according to nameplate
Loading after start	Fuse blowing. Fail to restart due to trip-off of automatic switch	Insufficient capacity of fuse	Replace fuse if wiring permits
		Overload	Lighten load
		High load at low voltage	Check circuit capacity and reduce load
	Overheating motor	Overload or intermittent overload	Lighten load
		Under-voltage	Check circuit capacity and power source
		Over-voltage	Check power source
		Ventilation duct clogged	Remove the foreign matter in the duct
		Ambient temperature exceeds 104°F (40°C)	Correct insulation class to B or F, or lower ambient temperature
		Friction between rotor and stator	Factory repair
		Fuse blown (Single-phase rotating)	Install the specified fuse
		Poor contact of circuit switches	Check and repair
		Poor contact of circuit starting switches	Check and repair
		Unbalanced three-phase voltage	Check circuit or consult power company

Industrial Electric Motors: By Mohammed H.A. Soliman

Kinds of Breakdown	Symptoms	Possible Causes	Remedies
Loading after start	Speed falls sharply	Voltage drop	Check circuit and power source
		Sudden overload	Check machine
		Single-phase rotating	Check circuit and repair
	Switch overheat	Insufficient capacity of switch	Replace switch
		High load	Lighten load
	Bearing overheating	High belt tension	Adjust belt tension
		Slack belt tension	Adjust belt tension
		Misalignment between motor and machine shafts	Re-align
		Over speed of bearing outer-ring	Adjust bracket
		High bearing noise	Replace the damaged bearing
Noise	Electromagnetic noise induced by electricity	Occurrence from its first operation	May be normal
		Sudden sharp noise and smoking	Short circuit of windings should be repaired at the factory
	Bearing noise	Noise of low shishi or Thru-Thru	May be normal
		Kala-Kala as a result of poor lubrication	Grease
		Kulo-Kulo as a result of poor lubrication	Clean bearing and grease
		Sa-Sa or larger noise	Replace the damaged bearing
	Mechanical noise caused by machinery	Loose belt sheave	Adjust key and lock the screw
		Loose coupling or skip	Adjust the position of couplings, lock key and screw
		Loose screw on fan cover	Lock fan cover screw tightly
		Fan rubbing	Adjust fan position
		Rubbing as a result of ingress of foreign matter	Clean motor interior and ventilation ducts
		Wind noise	Noise induced by air flowing through ventilation ducts
		Induced by conveyance machine	Repair machine
Vibration	Electromagnetic vibration	Short circuit of winding	Factory repair
		Open circuit of rotor	Factory repair
	Mechanical vibration	Unbalanced rotor	Factory repair
		Unbalanced fan	Factory repair
		Broken fan blade	Replace fan
		Unsymmetric centers between belt sheaves	Align central points
		Central points of couplings do not lie on the same level	Adjust the central points of couplings to the same level
		Improper mounting installation	Lock the mounting screws
		Motor mounting bed is not strong enough	Reinforce mounting bed

Problems of Poor Lubrication & Bearing Grease

	Properly Lubricated	Poorly lubricated
Sound	Only very weak sound could be heard from lace and retainer	Large noise from retainer or 2-pole motor's roller bearings.
Service life	The existence of oil film on various lubricated surface wear is greatly slowed down and long service life of bearings expected	Due to broken oil film, metal surfaces may contact each other. Excessive wear may occur and very short service life of bearings may be resulted in
Heat	From viscidity resistance.	From the friction of direct contacting of metal surfaces.
Damage		Damage and noise caused by metal dust. Burnt retainer. Deformed or broken rolling bodied and outer and inner wheels.
Vibration		Gaps between worn retainer, rollers and balls will get larger; deformation and vibration may be resulted in.
Damaged bearings will give a burnt motor		When bearings are damaged, rotor will come down and be in contact with stator. Heat from their friction will burn the windings.

Industrial Electric Motors: By Mohammed H.A. Soliman

Bearings Maintenance

If the instructions don't say otherwise, motors are usually greased with rolling bearings. All bearings are open except for close-type ball bearings. This helps prevent too much grease by making it easier to put in and take out the grease. Here are some tips for keeping things in good shape.

- ✓ Newly purchased motors need to be greased before starting them, or if they haven't been used for over 2 months.
- ✓ After the motors have started running. or once the motors are running. More oil needs to be added regularly, following the amounts listed on the plate with the name of the equipment.
- ✓ The leftover grease should be cleaned up promptly.

After stopping running for a long period (more than 2 months):

- As soon as you start the machine, you should add grease right away. The amount of medicine to be injected is displayed on the nameplate.
- The temperature around the bearings goes up after they start working.
- Loudness and sound quality of bearing noise.
- Loud sounds and shaking of the engine.
- Vibration of the bearing

Grease Supply

Keeping the right amount of grease in bearings is really important to keep them working well. The main reason for having grease available:

1. To keep the sliding surface greased.
2. To keep oil between moving parts to support weight and prevent damage.

If the grease stays in place and the surfaces slide smoothly, you won't hear any noise.
3. To clean away the old grease and dirt from the worn-out part.
4. Grease protects the bearing from rust, dust, and makes it quieter and less shaky.

Tips:

Ensure that you provide the right amount of grease at the specified times shown on the nameplate.

If your motor has been stopped for more than 2 months, a grease should be applied when the machines are turned on again.

When bearings get too much grease, they can get too hot and stay that way. When they don't get enough grease, it can't reach the inside of the bearings.

Intervals between grease supply: If the motor runs all day every day, how often will it need to be supplied? The information is shown on the name plate. The motor runs for 12 hours one day and 8 hours the next day, if someone works for 3 hours one day, it will be like working for 12 hours, every day, we decide how often to put grease in to make sure enough is being supplied effectively.

Table A Quantity and Interval of Grease Supply Ball bearing

Table A Quantity And Interval of Grease Supply Ball bearing

Bearing No.	First filling (1)(g)	Consecutive supplies (2) (g)	Interval between grease supplies for Motors runs 24 hours everyday (in days) (3)					
			2-p	4-p	6-p	8-p	10-p	12-p
6310,6210	50	30	120	180	180	180	180	180
6311,6211	100	30	120	180	180	180	180	180

Industrial Electric Motors: By Mohammed H.A. Soliman

Bearing	First filling (g)	Consecutive supplies (g)	4-p	6-p	8-p	10-p	12-p
6312,6212	100	30	120	180	180	180	180
6313,6213	100	30	120	180	180	180	180
6314,6214	200	50	80	180	180	180	180
6215,6214	200	50	-	180	180	180	180
6316,6216	200	50	-	180	180	180	180
6317,6217	200	50	-	180	180	180	180
6318,6218	300	50	-	180	180	180	180
6320,6220	400	80	-	120	180	180	180
6322,6222	600	80	-	120	180	180	180
6324,6224	600	80	-	120	180	180	180
6326,6226	1000	100	-	-	180	180	180

Table B Roller bearing

Bearing No.		First filling (1) (g)	Consecutive supplies (2) (g)	Interval between grease supplies for Motors runs 24 hours everyday (in days)(3)				
				4-p	6-p	8-p	10-p	12-p
NU3 ☐☐	14	100	50	180	180	180	180	180
	15	100	50	180	180	180	180	180
	16	100	50	180	180	180	180	180
NU2 ☐☐	17	200	50	120	180	180	180	180
	18	200	50	120	120	180	180	180
NU22 ☐☐	20	300	80	120	120	180	180	180
	22	300	80	120	120	180	180	180
	24	400	80	-	120	180	180	180
	26	600	100	-	120	180	180	180

Comments:

- The first time the grease is filled in the bearings after they are cleaned. One third will go in the bearing, and the rest will go in the bearing covers.
- Supply quantity is the amount of grease that needs to be given at regular intervals.
- If the motor runs 8 hours one day and 12 or 6 hours the next day, we will count it as running for 12 hours each day. The time shown in the tables can be multiplied by two.
- If the engine mentioned in the previous section. This machine has either 2 poles or 4-6 poles and uses big roller bearings. You shouldn't wait too long between putting grease in these bearings.
- Please don't try to make the time longer by giving more.

- For motors with 2 poles or 6 poles and big roller bearings, if they have been stopped for more than 2 months and started running again, you should add grease as per the instructions on the nameplate. If not, you might hear some noise and the bearings could get worn out or damaged in an unusual way.

Getting rid of grease

When the container holding the grease from the bearing gets full, the bearing may get too hot because the grease is thick and sticky. This might cause the grease to leak out. So, please open the cover and let the grease out on time.

Bearing Diagnosis

If you put in new grease and get rid of old grease the right way, the motor will work well. When engines don't work well because of bearing problems, here are some tips for figuring out what's wrong.

1. Loud sound coming from the bearings.

2. Bearings getting hot.

Temperature rise means how much hotter the bearings are compared to everything around them. It is usually shown as degrees Celsius (°C). The table shows the highest temperatures that are allowed to increase. Please check if anything is wrong if the temperature goes above 40 degrees Celsius.

Industrial Electric Motors: By Mohammed H.A. Soliman

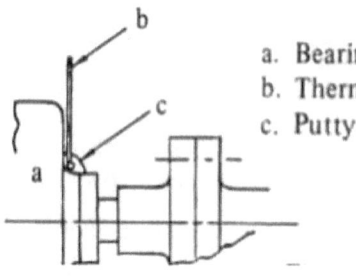

a. Bearing Bracket.
b. Thermometer.
c. Putty.

	Values of temperature rise (ambient temperature 40℃)	Readings from thermometers	Specifications
Cap grease	40°	80°	CNS 2934 JIS C4202 JEC 37 JEM 1020
Heat-resist grease	55°	95°	CNS 2934 JEC 37 JEM 1020

When the temperature changes while you are running, keep the following in mind:

- ✓ The ability of a new batch of grease to resist getting thick or sticky when it is put into or taken out of a machine.
- ✓ Not enough or worn-out grease.
- ✓ Grease that gets on bearing caps or retainers can get into the bearing and make it harder to move.

- ✓ The motor's temperature can change when the amount of work it has to do changes.

3. Variation of bearing vibration

As things like wheels and metal balls start to wear out, the spaces between them get bigger and the shaking gets worse. If the washer isn't tightened enough and there's not enough oil, the machine might shake and make a whistling noise. Balls, rollers, and retainer can get messed up and cause strong shaking. It's important to figure out what's causing the shaking and fix it. Defects on the surfaces where things roll can also make the shaking worse.

4. General appearance of discharged grease

Color and odor	- White mixture may be considered as air and water mixed into the grease when churned by the bearings. - The grease in a new machine may become dark due to the grindings from rough surfaces. In this case, check and see whether the metal dust has got into the bearing. - In addition to chemical change, dust, air bulbs and water may mix into grease and make it deteriorate, discolor and smell bad.
Hardness	- Insufficient supply of fresh grease, old grease will remain in the grease groove. Fresh grease will get into the groove when sufficiently supplied.
Foreign bodies	- Matters other than grease may be wrongly injected; dust may get in.

Industrial Electric Motors: By Mohammed H.A. Soliman

Choosing the Right Grease

Before you choose which grease to use, please think about these descriptions:

AVAILABILITY :	Choose the products of a world-wide supplier so that they will be always available .
TEMPERATURE :	Temperatures where in regular greases are serviceable range from -20° to 120℃ . Beyond this range , greases for low or high temperature should be adopted .
For high rpm motors and motors with bearing of large diameters	Harder grease has better compression strength while softer ones may give minimum noise and vibration and allow an easy operation in injection and discharge . (Silicon grease prohibited.)
LOAD- BEARABILITY :	Grease of good compression strength for heavily loaded operation (belt or gear transmission). Silicon grease prohibited .
Moisture-proof :	Na-grease or Ca-grease is recommended for motor installed in moist environment .
Viscidity :	Among different brands of grease with same hardness , the one with lower viscosity is recommended to minimize noise , vibration and temperature rise of bearings after greasing and to provide good lubrication during cold weather and easy operation in grease injection and discharge .
Serviceability :	The better you understand lubrication products the wiser decision you will make in grease selection .

Mixture of different grease

Lubricants with the same main ingredient and type, but with different thickness, can be mixed. If the grease you need to use is different from what is already in the bearings, you have no other option but to use it to fill them up. Inject a lot of the new grease to replace all of the old grease in the bearings.

Assembling and Disassembling

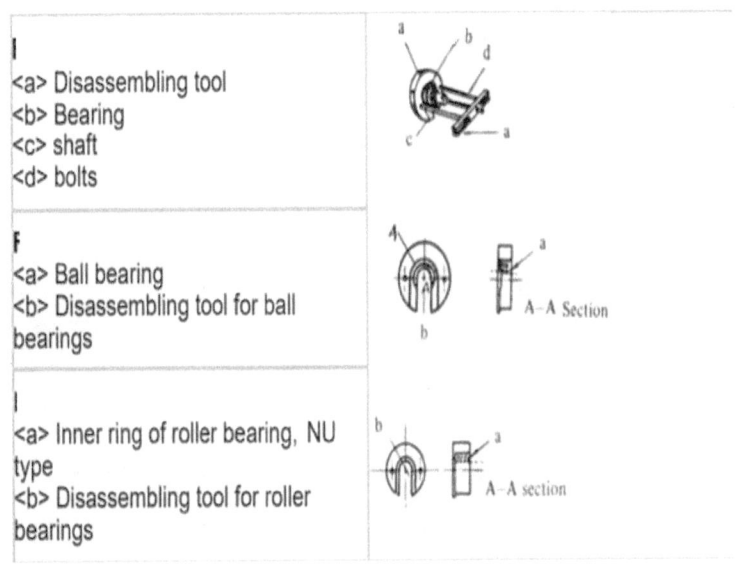

\<a\> Disassembling tool \<b\> Bearing \<c\> shaft \<d\> bolts	
\<a\> Ball bearing \<b\> Disassembling tool for ball bearings	
\<a\> Inner ring of roller bearing, NU type \<b\> Disassembling tool for roller bearings	

If the part taken off the shaft or object can be fixed, please clean it with machine oil or coal oil. Wrap it tightly in plastic to keep it dry and clean. Don't wash it before wrapping. Take apart the thing by following the steps in the pictures. The pulling force should be spread out evenly throughout the inner ring of the bearing.

Industrial Electric Motors: By Mohammed H.A. Soliman

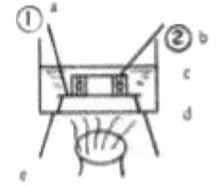

a. Propeller.
b. Thermometer.
c. Clear machine oil or coal oil.
d. Bearing support Stand.
e. Heat source.

Figure 16

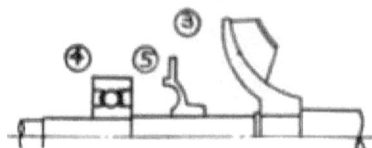

The steps below are for putting together bearings (fitting them tightly).

1. The bearing needs to be heated evenly and should not touch the container directly. Please mix the oil thoroughly.

2. To keep bearings from getting damaged and changed, they should not be heated above 120°C, even in one spot. Place the bearings in the oil and warm them up to 100 degrees Celsius. Then take out the bearings to make them fit better.

3. First, put the vanes and inner bearing cap together, if needed.

4. The side with the details or model number should be facing towards you.

5. After the fitting is done, the part should be pressed or hammered to make it tight against the surface of the bush. When you press or hit something, the force must be spread out on the inner ring as shown in the first picture. The second picture shows the wrong way to do it, which is not allowed at all.

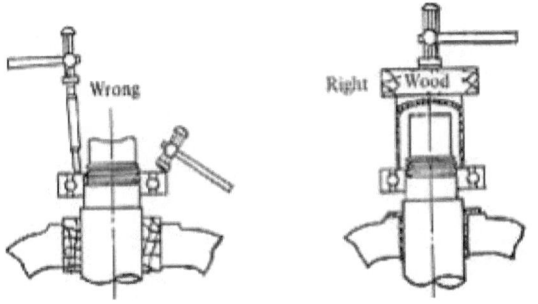

Also, remember these steps:

1. Check that the openings in the bearing caps and brackets where grease goes through are lined up with each other.

2. Fill the bearing and its caps with the main source of grease.

3. Sealant should be put into the gaps of outdoor motors.

Sealed Ball Bearing

Construction

Sealed ball bearings have plates that seal them on both sides, while open-type bearings do not have these plates. Due to the way it is built, the first one needs to be greased less often.

Knock-down test

A test should be done every 2 or 3 years. After the event

The motor has been used for two years. Use a stethoscope to check the bearings for strange noises. If they sound bad, replace them with new ones. The covers on sealed ball bearings can be removed to put in new grease. Before putting in new grease, make sure to clean off the old, worn-out grease completely with benzene or coal oil.

Electric Motor Reliability

Criticality of Electric Motors

The criticality of electric motors is based on its function inside the system. An electric motor that is supplying electricity to water pumps in a residential building is not as critical as the one supplying electricity to pumps or fans in a production process. An electric motor failure in nuclear industry can cause severe damage, safety risks or high losses.

The consequence of equipment failure like an electric motor or any other equipment can be catastrophic if it has great effect on safety (eg. supplying firefighting system), or production and costs (eg. Manufacturing plants).

Industrial Electric Motors: By Mohammed H.A. Soliman

Industrial Electric Motors: By Mohammed H.A. Soliman

What is risk management?

Is the way toward recognizing, evaluating and controlling dangers to an association's capital and profit. These dangers, or threats, could come from a wide assortment of sources, including monetary vulnerability, legitimate liabilities, vital administration mistakes, mishaps and catastrophic events.

Determination of risk level
The probability is the likelihood of an event occurring and the consequences, to which extent the project is affected by an event, are the impacts of risk. By combining the probability and impact, the Level of Risk can be determined.

Industrial Electric Motors: By Mohammed H.A. Soliman

Impact	Very Likely		Average		Very Low
No Impact	Priority 5	Priority 5	Priority 5	Priority 5	Priority 5
	Priority 3	Priority 3	Priority 3	Priority 5	Priority 5
Some Impact	Priority 2	Priority 2	Priority 2	Priority 5	Priority 5
	Priority 1	Priority 1	Priority 2	Priority 4	Priority 5
Disastrous Impact	Priority 1	Priority 1	Priority 2	Priority 4	Priority 5

Probability

You can always design this matrix for your system components to determine the risk priority value for each component.

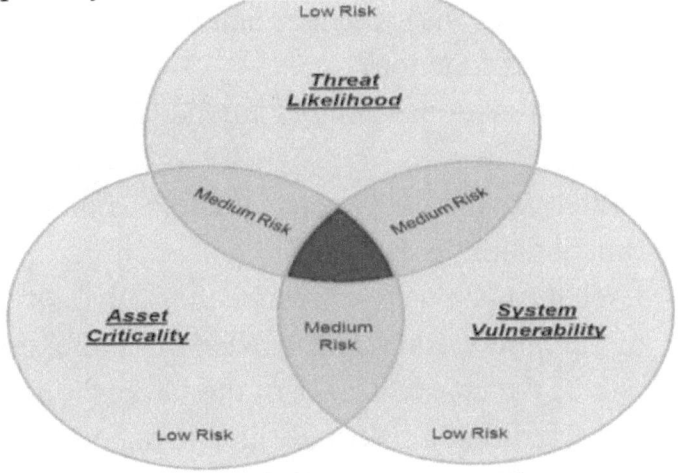

Knowing which component of the system is likelihood to fail is very useful in determining which strategy to use to prevent this failure. But first, you need to identify your priorities, which system component is likelihood to fail and need attention first, and if this component is critical to your business, your process or not.

Here are some few questions that help define the criticality of the system components:

The RCM analysis carefully considers the following questions:

- What does the system or equipment do; what is its function?
- What functional failures are likely to occur?
- What are the likely consequences of these functional failures?
- What can be done to reduce the probability of the failure, identify the onset of failure, or reduce the consequences of the failure?

In performing a risk analysis, the risk team uses a structured decision process to develop mitigating tasks for each failure mode identified during the analysis:

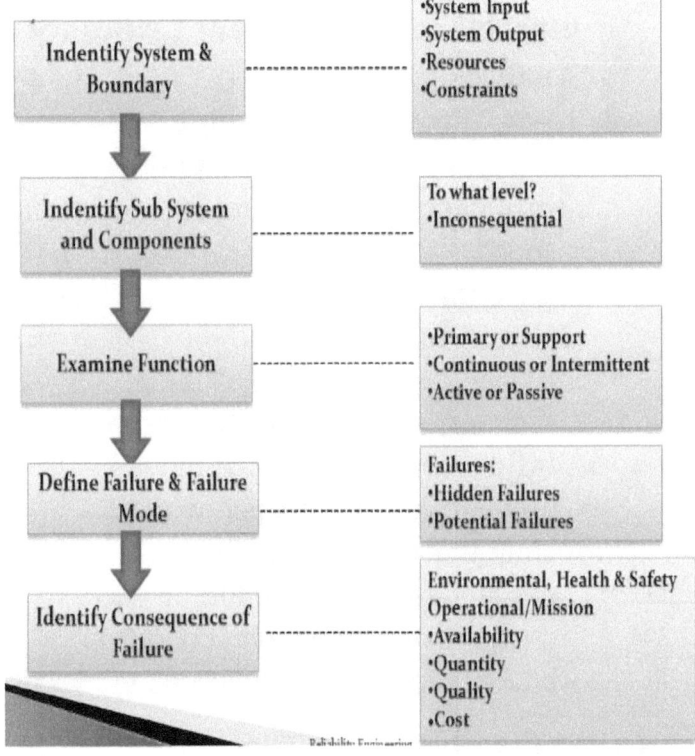

Asset Criticality Assessment

Importance of Equipment Criticality Analysis:
1. Influence the priority assignment of the Work Orders.
2. Influence the Work Orders execution speed.
3. Determine which Maintenance Class should come first.
4. Effect the scheduling of the preventive maintenance program.
5. Influence the priority of the preventive maintenance work.
6. Help in determining which maintenance approach to be used (corrective, preventive, condition based, proactive, risk based...etc.).

Criticality Measuring Principles

1. **Safety**
   - ✓ Safety equipment (equipment carrying peoples, firefighting system, and alike).

- ✓ Consequence of parts failure effect safety (elevator ropes broken, firefighting generator stopped…etc.).
- ✓ Gas piping leakage at any point poses a risk.

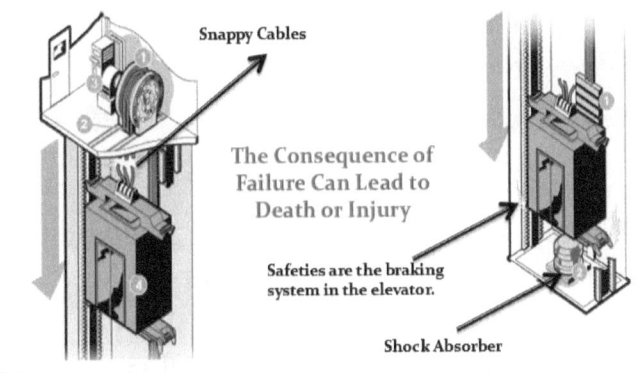

2. Production, Process
- ✓ Equipment breakdown affect the whole production line.
- ✓ Equipment breakdown affect partially the production line.
- ✓ Equipment failure has native effect on production quality.

The Consequence of Failure Can Lead to Process Shutdown, Major Losses, or Quality Problem.

Criticality decreases with redundancy in the system.

Criticality is influenced by the availability of standby equipment in a system but how much time does it take for the standby equipment to operate? And what is the effect of this on the production?

3. Time (Time=Money)
1 min production = How much?
1 Hr. production = How much?

If your equipment is classified as critical, ask yourself the following questions:
- What is the preventive maintenance program I have for it? Enough? Or not?
- How much time does it take to repair it in case of failure?

- Spare parts allocation? Available? Not available? If not available, how much does it take to allocated it from the vendor?
- Require special skills for repair? Is my team trained to repair it in a proper time?
- Do you have an emergency plan for it in case of accident or failure? Is your team aware of this plan and trained on it?

4. Money, Cost

Industrial Electric Motors: By Mohammed H.A. Soliman

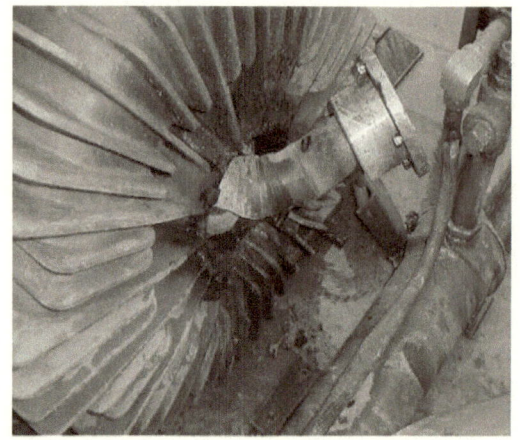

Equipment that fails in service can cost up to 10 times more to repair than the equipment repaired when predicted by condition monitoring.

Failure Consequence Criticality Classification

Class	Health/ Safety/ Environment	Production	Cost
High	•Potential for serious personnel injuries •Render safety critical system inoperable •Potential to fire in classified area •Potential for large pollution	Stop in production/significant reduce rate of production exceeding X hours	Substantial cost
Medium	•Potential for injuries requiring medical treatment •Limited effect on safety system •No potential for fire in classified areas •Potential for moderate pollution	Brief stop in production/reduced rate of production lasting less than X hours	Moderate cost
Low	•No potential for injuries •No potential for fire or effect on safety system •No potential for pollution (specify limit)	No effect on production within a defined period of time	Insignificant cost

Risk Class, SAFETY first!

Level	Description	Example
A	Major effect on HSE	Fire fighting generator
B	Major effect on process=high downtime cost	Furnace generator
C	Normal effect on HSE Normal effect on process Without Standby	
D	Normal effect on process Normal effect on HSE With Standby	

Failure Mode Effect Analysis FMEA as a Great Tool for Minimizing Risks of Failure and improving Reliability

An FMEA is a systematic method for identifying and preventing product and process problems before they occur. FMEAs are focused on preventing defects, enhancing safety and increasing customer satisfaction.

FMEAs are conducted in the product design or process development stages, although conducting an FMEA on existing products and processes can also yield substantial benefits.

What is the purpose of a FMEA?
Preventing the process and product problems before they occur is the purpose of Failure Mode Effect Analysis. Used in both the design and manufacturing process, they substantially reduce costs by identifying product and process improvement early in

the develop process when changes are relativity easy and inexpensive to make.

FMEA can provide the answer to many problems:

How can we prevent this problem from occurring again in the future?
How can we minimize the risk of this potential failure?
How can we produce an error-free product?
How can we reduce the warranty costs?
How can we improve the safety condition in the workplace?

Benefits of Failure Modes Effect Analysis "FMEA"

The object of an FMEA is to look for all of the ways a process or product can fail. A product failure occurs when the product does not function as it should or when it malfunctions in some way.

Contribute to improve design for product & process

- Higher reliability.
- Better Quality.

- Increase Safety.

Contribute to cost saving
- Decrease development time & redesign cost.
- Decrease warranty costs.
- Decrease wastes.

Contribute to continuous improvement

FMEA Applies to: System, Process, Design, and Service

FMEA helps manufacturing engineers control the process and eliminate errors during production, thus decreasing warranty costs and wastes.

Failure Mode

Any event which causes a functional failure. Another definition, ways in which product or process can fail are called failure modes. The FMEA is a way to identify the failures, effects, and risks within a process or product, and then eliminate or reduce them.

Industrial Electric Motors: By Mohammed H.A. Soliman

Example – Failure Modes for Common Equipment

There are many failure modes examples for different mechanical and electrical machines.

Electric Transformer

Failure Mode	Failure Mode	Failure Cause
Voltage regulation issue	Tap changers	Mechanical wear
High noise/ Corona	Cable insulation	Overload
High oil temp	Fins	Blockage
Oil leakage	Tank gasket damage	Wear

Vehicle

Failure Mode	Failure Mode	Failure Cause
Car won't start المحرك لا يدور	Fuel delivery pump طلمبه البنزين	Pump motor wear/fail عطل في الطلمبه
	Ignition coil الموبينا	Burn coil احتراق الموبينا
	Blockage in fuel filter انسداد في فلتر البنزين	Dirties in the tank شوائب في التنك
	Cam belt سير الكاتينه	Broken انقطاع
	Control unit وحده التحكم	Failed/Water got in عطل نتيجه تسرب مياه
Poor idling	Actuator sensor (IAVC)	Life time
	ECU Control unit	Failed
	Injectors الرشاشات	Blockage due to dirties in fuel انسداد نتيجه شوائب في البنزين
	Spark plugs البوجيهات	Wear تآكل أو تلف
	Fuel filter فلتر البنزين	Dirties causing blockage

Failure Mode	Failure Mode	Failure Cause
Car performance (acceleration problem)	Spark plugs	Wear
	Air Filter	Dirty
	Fuel filter	Blocked
	Fuel pump	Weak
	Speed sensor	Faulty
	ECU	Faulty
	Compression ratio	Worn cylinder head
		Cylinder head gasket
		Wear piston rings
		Internal wear

Industrial Electric Motors: By Mohammed H.A. Soliman

Generator

Failure Mode	Failure Cause
Engine is running, but no AC output is available	1. One of the circuit breakers is open. 2. Fault in generator. 3. Poor connection or defective cord set. 4. Connected device is bad.
Engine runs good at no-load but "bogs down" when loads are connected.	1. Short circuit in a connected load. 2. Engine speed is too slow. 3. Generator is overloaded. 4. Shorted generator circuit. 5. Clogged or dirty fuel filter.

Failure Mode	Failure Cause
Engine will not start; or starts and runs rough.	1. Start switch in off (O) position. 2. Fuel valve is in "Off" position. 3. Failed battery. 4. Low oil level. 5. Dirty air cleaner. 6. Clogged or dirty fuel filter. 7. Fill fuel tank. 8. Drain fuel tank and carburetor; fill with fresh fuel. 9. Spark plug wire not connected to spark plug. 10. Bad spark plug. 11. Water in fuel. 12. Flooded.

Failure Mode	Failure Cause
Engine is running, but no AC output is available	1. One of the circuit breakers is open. 2. Fault in generator. 3. Poor connection or defective cord set. 4. Connected device is bad.
Engine runs good at no-load but "bogs down" when loads are connected.	1. Short circuit in a connected load. 2. Engine speed is too slow. 3. Generator is overloaded. 4. Shorted generator circuit. 5. Clogged or dirty fuel filter.

Belt Conveyor

Failure Mode	Failure Cause
All Portions Of The Conveyor Belt Run To One Side At A Given Point On The Structure.	1. One Or More Idlers Immediately Preceding Trouble Not At Right Angles To The Direction Of Belt Travel. 2. Conveyor Frame Or Structure Crooked. One Or More Idler Stands Not Centered Under Belt. 3. Sticking Idlers. 4. Buildup Of Material On Idlers. 5. Structure Not Level And Belt Tends To Shift To Low Side.

Failure Mode	Failure Cause
A Particular Section Of The Conveyor Belt Runs To One Side At All Points On The Conveyor.	1. Belt Not Joined Squarely. 2. Bowed Belt 3. Worn Edge
Severe Wear On The Pulley Side Of The Conveyor Belt.	1. Slippage On Drive Pulley 2. Material Spills Between Belt And Pulley. 3. Material Build Up At Loading Point Until Belt Is Dragging 4. Sticking Idlers 5. Excessive Tilt To Trough in Idlers 6. Bolt Heads Protruding Above Lagging 7. Bottom Cover Too Thin.

Industrial Electric Motors: By Mohammed H.A. Soliman

Failure Mode	Cause
The Conveyor Belt Runs To One Side For Long Distance Along The Bed.	Load Being Placed On Belt Off Center. Conveyor Frame Or Structure Crooked.
Fasteners Pull Out	1. Wrong Type Of Fastener Or Fasteners Not Tight 2. Tension To High 3. Heat 4. Tandem Drive Poorly Compensated 5. Inadequate Convex Curve Radius
Drive Pulley Spins	1. Belt Tension. 2. Drive Pulley lag
Conveyor belt breaks	1. Belt Overloaded 2. Splice Failure 3. Belt Wear From Age
Belt Moves Sideways, After Making A Complete Cycle	Belt Was Not Squared When Spliced.

Failure Mode	Cause
Excessive material carry back on the belt's top cover causes build up on the snub pulley and return idlers	No belt scraper Bad belt scraper Bad lagging
Material splicing separated	Material buildup migrate and grind into the top cover and into small imperfections in a belt splice

Industrial Electric Motors: By Mohammed H.A. Soliman

Failure modes at Spectrum Vibration

Failure Mode: Frequency in terms of RPM	Most likely causes	Other possible causes and remarks
1x RPM	Unbalance	1) Eccentric journals, gears or pulleys 2) Misalignment or bent shaft - if high axial vibration 3) Resonance 4) Reciprocating forces 5) Electrical problems
2x RPM	Mechanical Looseness	1) Misalignment if high axial vibration 2) Reciprocating forces 3) Resonance 4) Bad belts if 2x RPM of belt
3x RPM	Misalignment	Usually a combination of misalignment and excessive axial clearances (looseness)

Industrial Electric Motors: By Mohammed H.A. Soliman

Less than 1x RPM	Oil whirl (less than ½ RPM)	1) Bad drive belts 2) Background vibration 3) Sub-harmonic resonance
Synchronous (A.C. Line Frequency)	Electrical Problems	Common electrical problems include broken rotor bars, eccentric rotor, unbalanced phases in poly-phase systems, unequal air gap.
2x Synch. Frequency	Torque pulses	Rare as a problem unless resonance is excited
Many times RPM (harmonically related freq.)	Bad gears Aerodynamic forces Hydraulic forces Mechanical looseness Reciprocating forces	Gear teeth times RPM of bad gear Number of fan blades times RPM Number of impeller vanes times RPM May occur at 2,3,4 and sometimes higher harmonics if severe looseness
High frequency (not harmonically related)	Bad anti-friction bearings	1) Bearing vibration may be unsteady- amplitude and frequency 2) Cavitations, recirculation and flow turbulence cause random, high frequency vibration. 3) Improper lubrication of journal bearings (friction excited vibration) 4) Rubbing

Industrial Electric Motors: By Mohammed H.A. Soliman

Failure Modes at Ultrasound Detector

Ultrasound Noise (Symptom) - Failure Mode	Failure Cause
Steady regular buzzing or frying sound when measuring high-medium voltage devices	Corona
A buzzing and intermittent crackling sound when measuring high-medium voltage devices	Arcing
A violent sound of an electrical arc with an abrupt start and stop when measuring high-medium voltage devices	Tracking
High frequency spectrum when measuring the bearing	Bad bearing lubrication
Friction sound when measuring compressors reciprocating valves	Valve wear

Example Failure Mode, Effects, and Causes:

Ex.1 Centrifugal Fan

Failure mode	Failure Effect	Failure Effect (System)	Failure Effect (End)	Failure cause Level 1	Root cause
Fan operate with high vibration level	Equipment damage/breakdown	Unexpected plant shutdown	Major production losses	Bearing fails	Poor Maint
	Equipment damage/breakdown	Unexpected plant shutdown	Major production losses	Housing wear	Poor Maint
	Equipment damage/breakdown	Unexpected plant shutdown	Major production losses	Unbalance fan blade	Poor Maint
	Equipment damage/breakdown	Unexpected plant shutdown	Major production losses	Looseness in foundation	Poor Maint
	Equipment damage/breakdown	Unexpected plant shutdown	Major production losses	Shaft wear	Poor Maint

Industrial Electric Motors: By Mohammed H.A. Soliman

Ex.2 Transformer

Item name	Failure mode	Failure Effect (local)	Failure Effect (System)	Failure cause	Failure Cause	Root cause
Oil	1.Short circuit in transformer	Functional stop	Production losses	Particles in the oil	Overheated	Bad Maintenance
		Functional stop	Production losses	Water in the oil	Overheated	Bad Maintenance
					Aging	
Tap Changes	2-Can't change voltage level	Functional stop	Production losses	Mechanical damage	Wear	Life time/ maintenance

Ex.3 Water System

Function	Functional failure/failure modes	Causes
Provide water to the industrial process	Total loss of pressure, volume & flow	Pump failed Motor failed Valve out of position

Electric Motor

Function	Functional failure/failure modes	Causes
Drive the water pump	Burn out	Circuit Breaker tripped Bearing seized Insulation Rotor Insulation Stator

Motor Bearing

Failure mode	Failure Cause	Sources of failure/causes	Causes
Bearing seized, this include bearing, seals, lubrication	Lubrication	Contamination	Supply dirty Sealing failed
		Wrong type	Procedure wrong Supply information wrong
		Tool little	Human error Procedure error
		Too much	Human error Procedure error

Final Table

Failure effect			Severity			Causes	Root Cause	Occurrence	Current fault detection methods	Detection	RPN	Actions
Local	sys	end	S	A	C	Seal failed	Seal failed					
Motor shutdown	System shutdown	TPL				Procedure wrong	Lack of training					
						Human error						
						Human error						

Overview on the FMEA Process and How to Conduct It

There are twelve steps for a successful FMEA process.

1. Select a high-risk process, then follow these steps.
2. Review the process: this step usually involves a carefully selected team that includes people with various job responsibilities and levels of experiences. The purpose of an FMEA team is to bring a variety of perspectives and experiences to the project.
3. Breakdown the system into components and sub-components.
4. Brainstorm potential failure modes.
5. List potential effects of each failure mode.
6. Assign a severity ranking for each effect.
7. Assign an occurrence ranking for each failure mode.
8. Assign a detection ranking for each failure mode.
9. Calculate the risk priority number (RPN) for each effect.

10. Prioritize the failure modes for action using RPN.
11. Take action to eliminate or reduce the high-risk failure modes.
12. Calculate the resulting RPN as the failure modes are reduced or eliminated.

FMEA Working Sheet
Component/Item Name:
Function

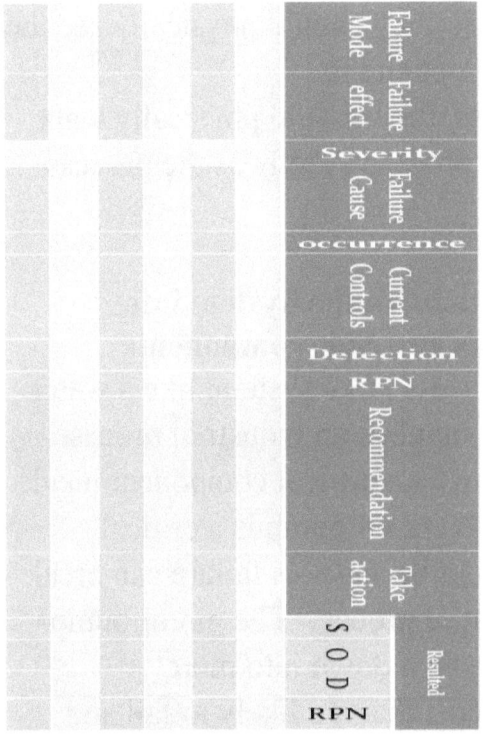

Step.2 Review the Process or Product

If the team is considering a product, they should review the engineering drawing of the product.

If the team considering a process, they should review the operation flowchart.

This is to ensure that everyone has the same understanding about the process or product.

For a product, they should physically see the product and operate it.
For a process, they should physically walk through the process exactly as the process flows.

Step.3 Breakdown the System into Components and Sub-components

If the system is a large system, like a water system that supplies an industrial process, the pump can be a critical component inside the system. A motor pump is a critical subcomponent because its failure can break down the entire process. The motor pump should be broken down into more subcomponents that are likely to fail and will affect the system, such as the motor's bearings and the rotor shaft. The FMEA will be used to prevent the probability of failure for each component or subcomponent.

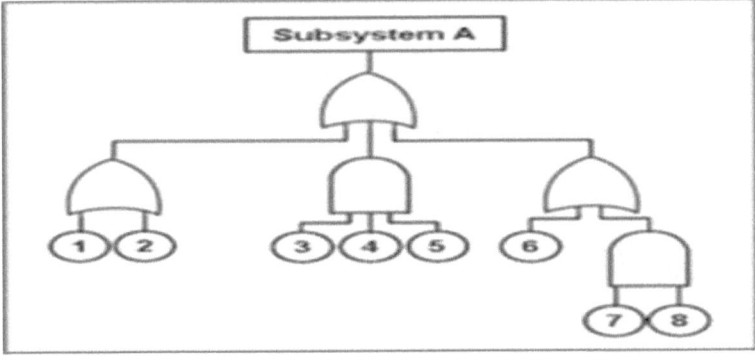

Step.4 Brain Storm Potential Failure Modes

When everybody in the group has a comprehension about the item or the cycle, colleagues should start contemplating the potential failure modes that could influence the process or the item quality. Zeroing in ought to be on the various components (individuals, material, management, strategy… and so on). When the conceptualizing is finished, the thoughts ought to be sorted out by gathering them into like classes. There are numerous approaches to assemble failure modes, they can be gathered by kind of failure (electrical, mechanical, client made). Where on the item or cycle the failure happens.

Main Rules of Brainstorm:

Try not to remark on, judge or study thoughts at the time they are advertised. Empower inventive and odd thoughts. The objective is to wind up with an enormous number of thoughts; and assess thoughts later. Every thought ought to be recorded and numbered precisely as offered, on a flip outline.

Expect to generate at least 50 to 60 concepts in a 30-minute brainstorming session.

Industrial Electric Motors: By Mohammed H.A. Soliman

Failure Mode & Effect Analysis FMEA

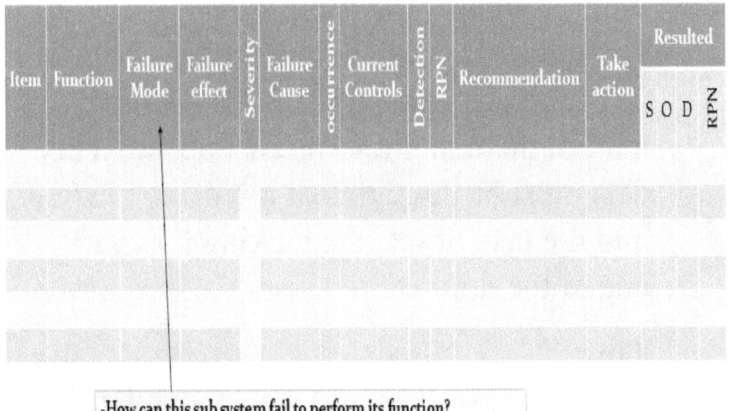

- How can this sub system fail to perform its function?
- The Way the failure occurred
- What will the operator see?

Step.5 List Potential Effects for Each Failure Mode

For a portion of the failure modes, there might be one impact, while for different modes, there might be a few impacts. This data must be through that it will take care of into the task of the risk ranking for every one of the failures.

Tips:

One failure mode could have several effects. For example, an electrical cutoff in the home could stop the refrigerator and damage food or prevent you from doing work on the computer.

Several failure modes could have one effect. A dead car battery or tire failure has the same effect on your vehicle – it will be difficult to make it to work on time with such a failure early in the morning.

The team must determine the end-effect each failure mode has on the system or the process. This means examining how each failure affects the entire system, the facility or the other connected processes.

Failure Mode & Effect Analysis FMEA

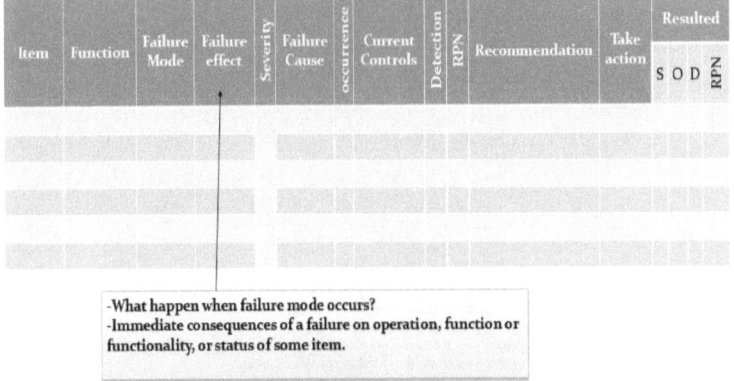

- What happen when failure mode occurs?
- Immediate consequences of a failure on operation, function or functionality, or status of some item.

Steps 6-8 Assign Severity, Occurrence, and Detection Rankings

Each of these three rankings is based on 10-point scale, with 1 being the lowest ranking, and 10 the highest.

Industrial Electric Motors: By Mohammed H.A. Soliman

Failure Mode & Effect Analysis FMEA

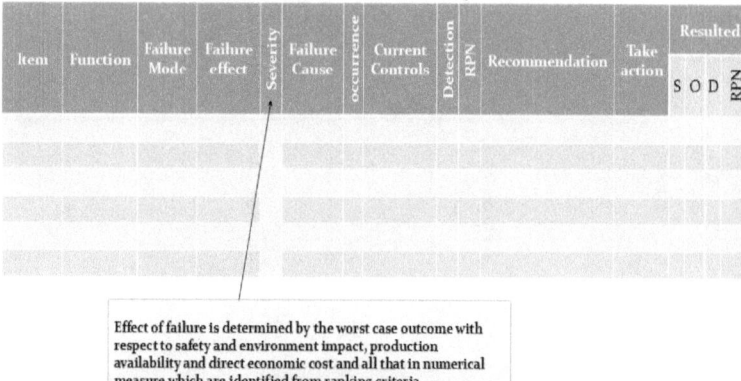

Effect of failure is determined by the worst case outcome with respect to safety and environment impact, production availability and direct economic cost and all that in numerical measure which are identified from ranking criteria

Failure Mode & Effect Analysis FMEA

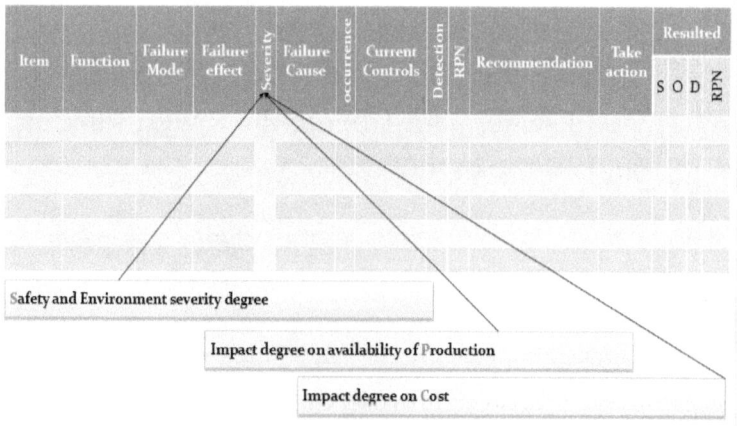

Safety and Environment severity degree

Impact degree on availability of Production

Impact degree on Cost

Severity Ranking Criteria

Description of Failure Effect	Effect	Ranking
No reason to expect failure to have any effect on Safety, Health, Environment or Mission.	None	1
Minor disruption of production. Repair of failure can be accomplished during trouble call.	Very Low	2
Minor disruption of production. Repair of failure may be longer than trouble call but does not delay Mission.	Low	3
Moderate disruption of production. Some portion too of the production process may be delayed.	Low to Moderate	4
Moderate disruption of production. The production process will be delayed.	Moderate	5
Moderate disruption of production. Some portion of production function is lost. Moderate delay in to High restoring function.	Moderate to High	6
High disruption of production. Some portion of production function is lost. Significant delay in restoring function.	High	7
High disruption of production. All of production function is lost. Significant delay in restoring High function.	Very High	8
Potential Safety, Health or Environmental issue. Failure will occur with warning.	Hazard	9
Potential Safety, Health or Environmental issue. Failure will occur without warning.	Hazard	10

Step.7 Assign an Occurrence Ranking for each Failure Mode

The best technique for deciding the occurrence ranking is to utilize real information from the process. This might be as failure history. At the point when real failure information is not accessible, the group must gauge how frequently a failure mode may happen, the group can improve gauge on how likely a failure mode is to happen and at what recurrence by knowing the expected reason for failure. When the potential causes have been distinguished for the entirety of the failure modes, an occurrence ranking can be appointed regardless of whether the failure information are not existed.

Failure Mode & Effect Analysis FMEA

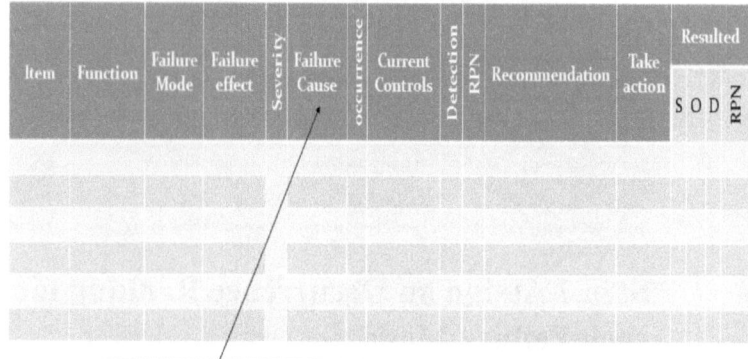

For each failure mode there may be several failure causes. Assign a Cause for each failure mode.
Select only potential failure to get failure causes.
Use Why Why Technique to get the root causes.
Identifying the failure cause can be the second option to determine the occurrence if no data is available in the form of failure logs.

Industrial Electric Motors: By Mohammed H.A. Soliman

Failure Mode & Effect Analysis FMEA

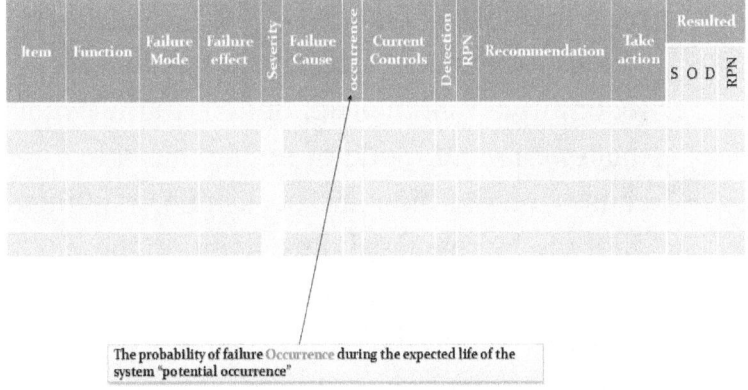

The probability of failure Occurrence during the expected life of the system "potential occurrence"

Occurrence Ranking Criteria

Rank	Freq	Description
1	1/10,000	Remote probability of occurrence; unreasonable to expect failure to occur
2	1/5,000	Low failure rate; similar to past design that has, in the past, had low failure rates for given volume or load
3	1/2,000	Low failure rate; similar to past design that has, in the past, had low failure rates for given volume or load
4	1/1000	Occasional failure rate; similar to past design that has, in the past, had similar failure rates for given volume or load
5	1/500	Moderate failure rate; similar to past design that has, in the past, had moderate failure rates for given volume or load
6	1/200	Moderate failure rate; similar to past design that has, in the past, had moderate failure rates for given volume or load
7	1/100	High failure rate; similar to past design that has, in the past, had high failure rates that have caused problems
8	1/50	High failure rate; similar to past design that has, in the past, had high failure rates that have caused problems
9	1/20	Very High failure rate; almost certain to cause Problems
10	1/10	Very High failure rate; almost certain to cause Problems

Operating hours based on the automotive industry benchmark.
Ranking can be determined based on historical data or similar system benchmarking

Step.8 Assign a Detection Ranking for each Failure Mode and/or Effect

To start with, the current control ought to be recorded for the entirety of the failure modes, or impacts, and afterward the recognition rankings appointed. In the event that one failure mode or impact has a few causes, recognition and occurrence rankings ought to be relegated dependent on these causes. At the point when potential causes are disposed of, the danger of failure is brought down.

In case the application is for an equipment maintenance, current control methods can be the current preventive maintenance program and or the current detection methods (condition monitoring program).

Industrial Electric Motors: By Mohammed H.A. Soliman

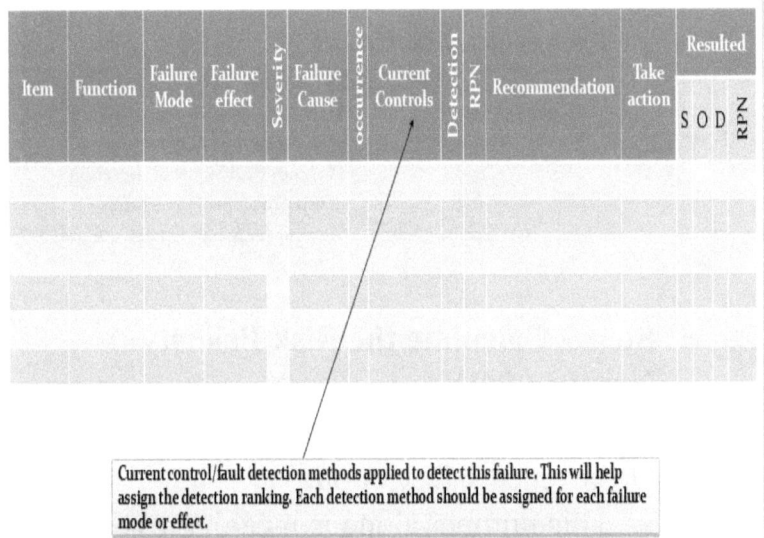

Current control/fault detection methods applied to detect this failure. This will help assign the detection ranking. Each detection method should be assigned for each failure mode or effect.

Failure Mode & Effect Analysis FMEA

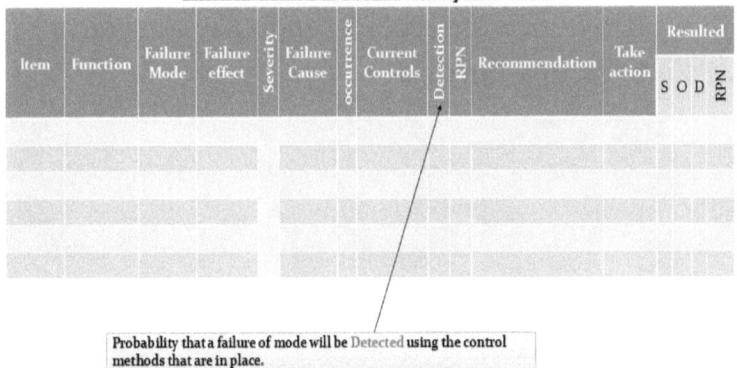

Probability that a failure of mode will be Detected using the control methods that are in place.

Detection Ranking Criteria

Rank	Description
1-2	Very high probability of detection
3-4	High probability of detection
5-7	Moderate probability of detection
8-9	Low probability of detection
10	Very low probability of detection

Step.9 Calculate the Risk Priority Number RPN

Risk Priority number= Severity x Occurrence x Detection.

This number alone is meaningless because each FMEA has a different number of failure modes and effects. However, it can serve as a gauge to compare the revised RPN once the recommended actions have been instituted.

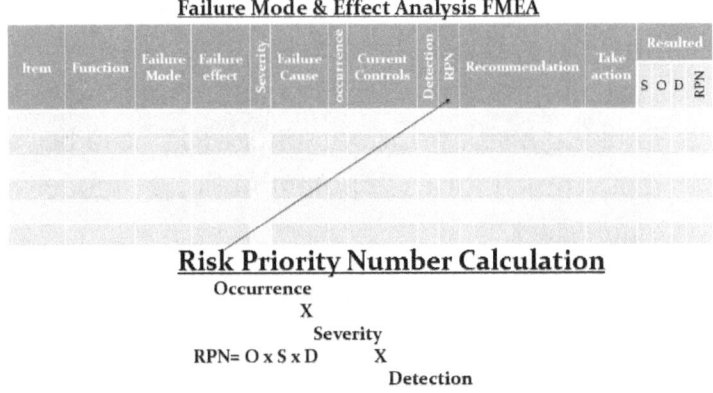

What is RPN?

The Risk Priority Number (RPN) methodology is a technique for analyzing the risk associated with potential problems identified during a Failure Mode and Effects Analysis (FMEA).

RPN Calculation Benefits:

- Contribute in Risk Assessment.
- Compare components to determine priority for corrective action. Components with higher RPN are given more attention.
- Assessing the risk priority number.

Each potential failure mode or effect is rated in each of these three factors on a scale ranging from 1 to 10. By multiplying the ranking a risk priority number RPN can be determined for each potential failure mode and effect.

The RPN will range from 1 to 1000 for each failure mode. It is used to rank the need for corrective action. Those failure modes with the highest RPN number should be attended first. Although the special attention should be given when the severity ranking is high from (9 to 10) regardless of the RPN.

Once a corrective action is takes, a new RPN is determined. This new RPN is called the resulting RPN.

Step.10 Prioritize the Failure Modes for Action

Failure modes ought to be organized by positioning them all together, from the most elevated danger need number to the least. Odds are that you will find that the standard 80/20 principle applied with the RPNs. A Pareto chart should be created.

The group should now choose which thing to work for. Typically, it assists with setting a cutoff RPN (cutoff point), where any failure modes with a RPN over that point are taken care of. Those beneath the cutoff are disregarded until further notice.

Tips:

High-risk numbers should be given attention first; then you can pay attention to the severity rankings. Thus, if several failure modes have the same risk priority number, that failure mode with the highest severity should be given more priority. If severity number is the same, those failures with higher occurrence should be given more priority and so on.

Failure Mode & Effect Analysis FMEA

Item	Function	Failure Mode	Failure effect	Severity	Failure Cause	occurrence	Current Controls	Detection	RPN	Recommendation	Take action	Resulted S O D	RPN

Appropriate maintenance action, appropriate maintenance task

Corrective actions may include: reduce the severity of occurrence, or increase the detection probability

Step.11 Take Actions to Eliminate or Reduce the High-Risk Failure Modes

This is organized using the problems-solving approaches and implement actions to reduce or eliminate the high-risk failure modes.

Often the easiest way to make an improvement to the product or process is to increase the detectability of the failure, thus lowering the detection rate.

Increase the detection rate can be done though assigning a schedule PM action, use a proper condition monitoring program or consider a mistake proofing method in the design. For example, ac computer software will automatically warn in case of low disk space.

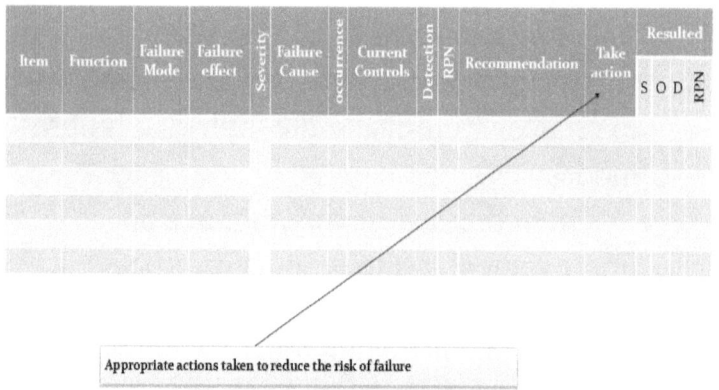

Appropriate actions taken to reduce the risk of failure

Step.12 Calculate the Risk Priority Number RPN as the High Risk is Removed

When moves have been made to lessen the danger need number, another positioning for the seriousness, event, and discovery ought to be determined. What's more, a subsequent RPN is determined. Desire is in any event 50 rate decrease in RPN with the FMEA approach.

There will consistently be a potential for failure modes to happen. The inquiry the organization must pose is how much relative danger the group is eager to take. That answer may depend on the business and the reality of the disappointment. For instance, in the atomic business, there is a little edge for mistakes, they can't hazard a calamity happening. In different enterprises, it might be worthy to face the high challenge.

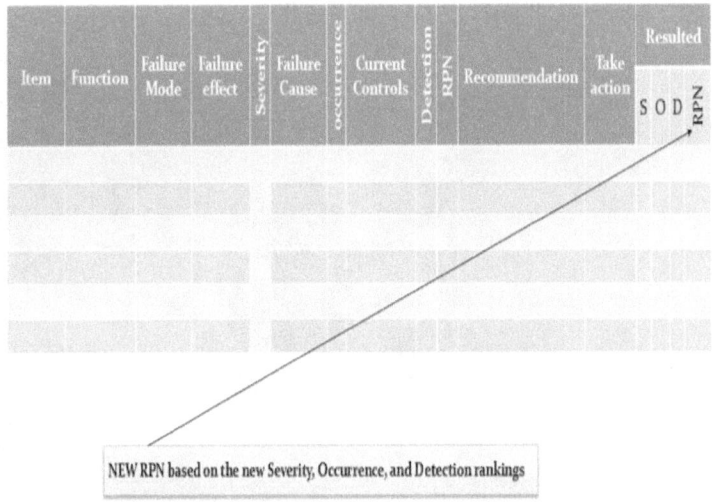

NEW RPN based on the new Severity, Occurrence, and Detection rankings

Industrial Electric Motors: By Mohammed H.A. Soliman

Improving the Reliability of a Pump Station System

After analyzing the system, the team came to the most common failures are coming from an electric motor exists in this pump station.

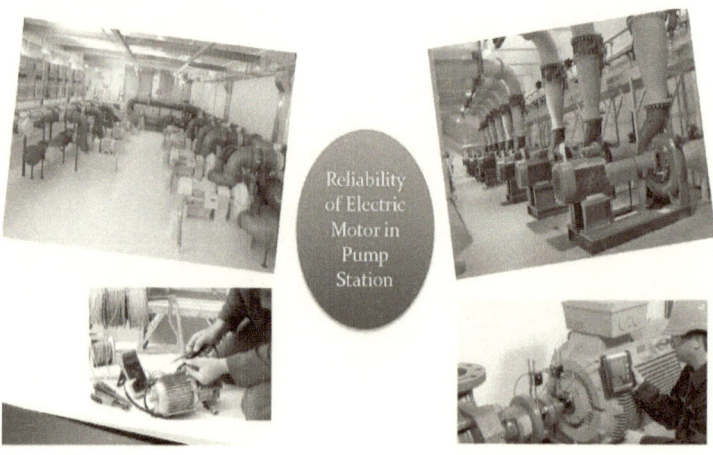

Industrial Electric Motors: By Mohammed H.A. Soliman

Breaking down the components of the electrical motor

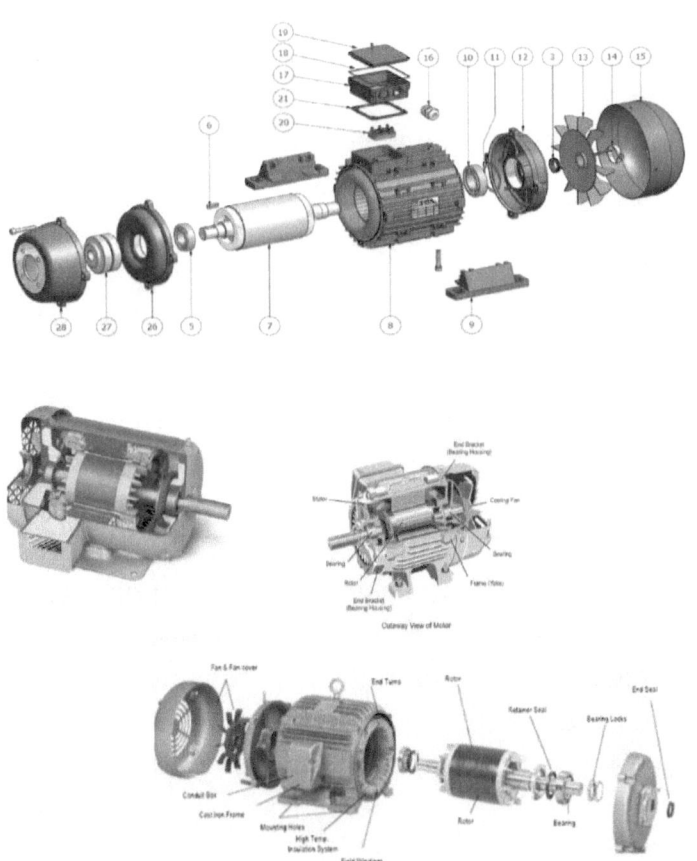

Equipment Information

Equipment Type   : DC Electric Motor

Technical Specs   : 50KW

Function   : Supply electricity to the water pump

System   : Water pump station, supply water for the industrial process

Availability of standby system: No

The electric motor is considered critical because a failure causes high production losses. The pump supplies the industrial process with the water required for the crystal manufacturing. Using a standby was not possible at the moment because it will require a change in the design which is very costly.

Industrial Electric Motors: By Mohammed H.A. Soliman

Electric Motor Failure Components

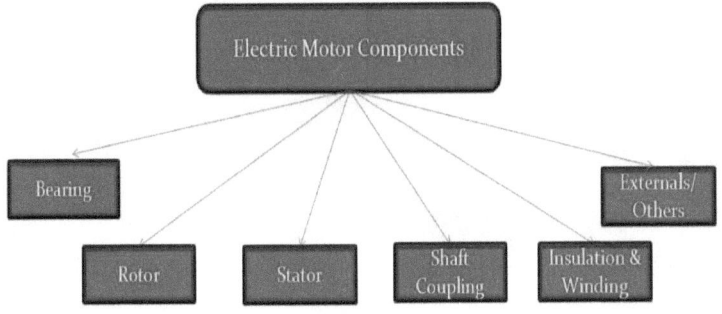

Electric motor sub-components

Electric Motor Failures

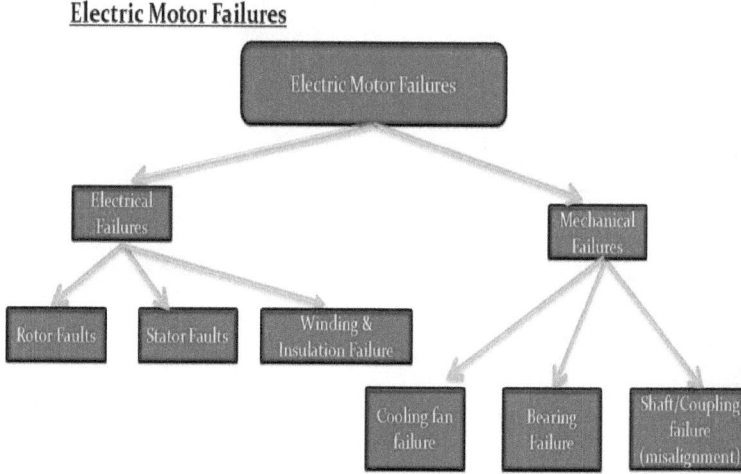

- ✓ You can group failures by mechanical, electrical…etc.
- ✓ More than 40% of motor failures are bearing failures.
- ✓ More than 25% of motor failures are stator failures.

Current Control/Prevention methods

PM Activity	Component/Item	PM Level
Check air gap between the rotor and stator with feeler gages	Rotor	Annually
	Stator	Annually
Inspect for misalignment	Coupling	Annually
Inspect for excessive wear. Check for proper type, hardness, conductivity, and fit in brush holders. Check holder spring pressure with a small scale. In most instances, pressure should be 2 to 2 1/2 lbs per sq in. of brush cross-sectional area	Brushes and commutators	Annually
Visual Inspection	Motor mounts, bolts, nuts, and tightening	Annually
Clean / Blow out dirties with compressed air. Remove any dust, chemicals, grease or dirt	Motor body and the cooling fan	Annually

Industrial Electric Motors: By Mohammed H.A. Soliman

Failure Log History

Working Condition= 24 hrs.

Failure type	Causes	Frequency (5 yrs)
Overheating & overloading	Bearing damage	6
Poor output	Motor inefficiency due to stator problems	2
High vibration	misalignment	2

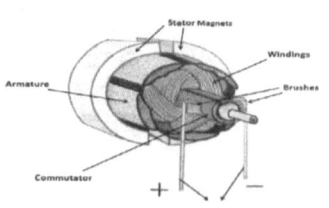

Failure Analysis

Component Name & Function: Bearing, reduce friction of the rotating shaft

Failure Mode	Failure Effect	Severity	Failure Causes	Failure Cause	Failure Causes	Occurrence	Current control methods	Detection	RPN
High vibration. Bearing Failed/ Seized	overload & overheat	8	Improper lubrication or grease	In correct type of lubricant		1	NA	8	64
				Wrong procedure	Lack of training	1		10	80
				In sufficient lubricant	Lack of maintenance	5		8	320
	High repair cost due to possible shaft damage		Improper mounting	Lack of tools		1	NA	1	8
				Lack of training	No standard	2		10	160
			Parts damage	Aging	Lack of maintenance		Annual inspection of shaft/coupling misalignment		
			Shaft misalignment	Lack of maintenance		2		8	128

Recommendation	Take actions	Result			
		S	O	D	RPN
1. Monitor bearing health condition with vibration and thermograph 2. Use standard procedures for bearing mounting and lubrication	Monitor vibration monthly, use standard work process and train workers them. Keep auditing the maintenance procedures	8	1	2	16
		8	1	2	16

Industrial Electric Motors: By Mohammed H.A. Soliman

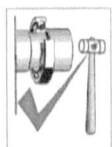
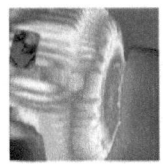

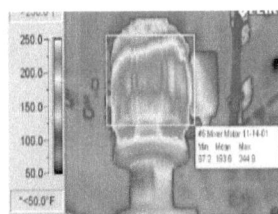

Bearing damage due to incorrect fitting

Component Name & Function: Coupling & shaft, transmit the movement

Failure Mode	Failure Effect	Severity	Failure cause	Failure Cause	Failure Cause	Failure Cause	Occurrence	Current controls	Detection	RPM
Misalignment	Equipment shutdown to avoid bearing damage and expensive repairs	8	Physical damage	Wear	Aging	Bad maintenance	1	Visual inspection	6	48
			Improper manufact	Bad type/material			1		10	80
			Improper installation	Bad Maint			1	Misalignment check annually	8	64
			Corrosion	Bad Maint			1		10	80

Recommendation	Take actions	Result			
		S	O	D	RPN

Industrial Electric Motors: By Mohammed H.A. Soliman

Flexible Coupling

Self Aligning Ball Bearing

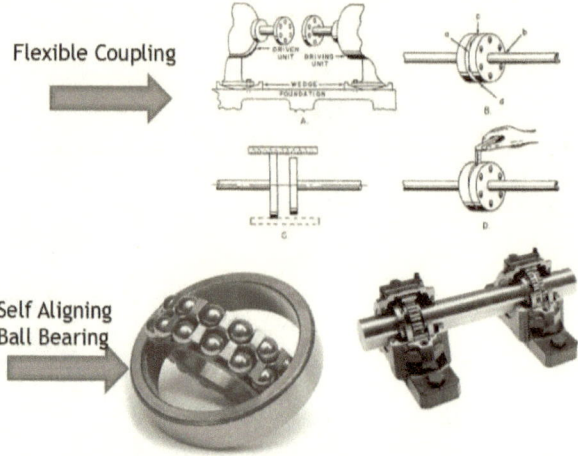

Industrial Electric Motors: By Mohammed H.A. Soliman

Precision Maintenance "Shaft Alignment"

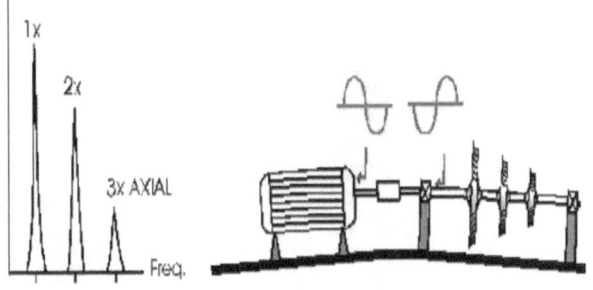

Angular misalignment detection using vibration analysis

Industrial Electric Motors: By Mohammed H.A. Soliman

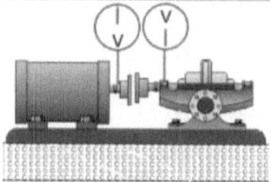

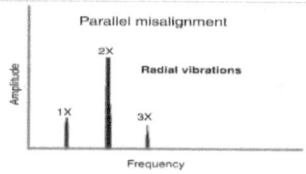

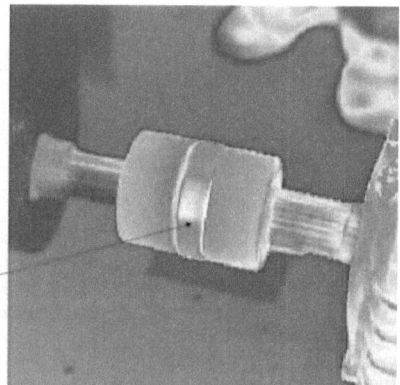

Coupling problem

Industrial Electric Motors: By Mohammed H.A. Soliman

Name & Function: Stator, generate electricity, carry current, retain armature

Failure Mode	Failure Effect	Severity	Failure cause	Failure Cause	Failure Cause	Occurrence	Current Controls	Detection	RPN
Stator defect	Motor inefficiency, High cost to repair	8	Eccentricity	High temp	Wear/Aging/Lack of maintenance	5	Check air gap between the rotor and stator with feeler gages	10	400
				Corrosion					
				High temp					
				Contamination					
			Short lamination	High temp					
				Corrosion					
				High temp					
				Contamination				10	400
			Loose iron	High temp					
				Corrosion					
				High temp					
				Contamination					

Recommendation	Take actions	Result			
		S	O	D	RPN
Monitor motor stator condition & temp with a proper method	Use vibration analysis & infrared thermograph analysis	8	1	1	8
		8	1	1	8

Stator defects

Eccentricity
Short lamination
Loose iron

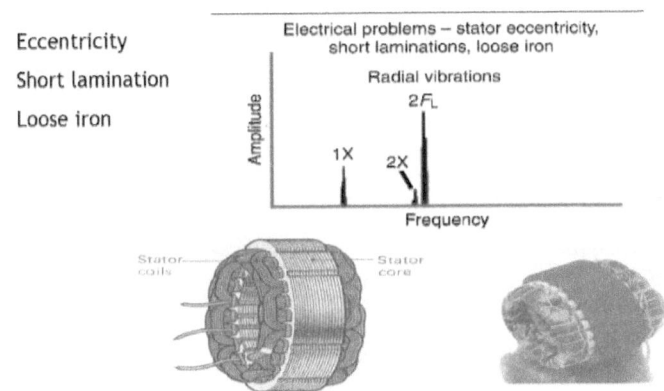

Industrial Electric Motors: By Mohammed H.A. Soliman

Stator Coil

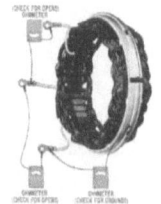

Charging stator coils

Minor core damage

Core damage as it appears in infrared

A D.C. motor consists of a rectangular coil made of insulated copper wire wound on a soft iron core. This coil wound on the soft iron core forms the armature. The coil is mounted on an axle and is placed between the cylindrical concave poles of a magnet.

Industrial Electric Motors: By Mohammed H.A. Soliman

Component Name & function: Rotor, is the moving part of the motor

Failure Mode	Failure Effect	Severity	Failure cause	Failure Cause	Failure Case	Cause	Occurrence	Current Controls	Detection	RPN
Rotor defect	Bearing damage, motor re build, high repair cost	8	Eccentric rotor	Imbalance	Wear	Aging	1	Checking air gap between the rotor and stator with feeler gages	10	80
				Thermal stress						
				Assembly problem	Bad maintenance					
				Soft foot or poor base						
			Broken rotor bars	Imbalance	Wear	Aging	1		10	80
				Thermal stress						
				Assembly problem	Bad maintenance					
				Soft foot or poor base						

Recommendation	Take actions	Result			
		S	O	D	RPN
Monitor rotor condition with a proper method	Use vibration analysis to detect rotor defects	4	1	2	8
		4	1	2	8

The rotor is a moving component of an electromagnetic system in the electric motor, electric generator, or alternator. Its rotation is due to the interaction between the windings and magnetic fields which produces a torque about the rotor's axis. When the coil rotates, the shaft attached to it also rotates and thus it is able to do mechanical work.

Industrial Electric Motors: By Mohammed H.A. Soliman

Damaged motor shaft

Detection of Different Electric Problems

The following are some terms that will be required to understand vibrations due to electrical problems:

F_L = electrical line frequency (50/60 Hz)

$$F_s = \text{slip frequency} = \frac{2 \times F_L}{P} - \text{rpm}$$

F_p = pole pass frequency = $F_s \times P$

P = number of poles.

Rotor problems

1. Broken rotor bars
2. Open or shorted rotor wind
3. Bowed rotor
4. Eccentric rotor

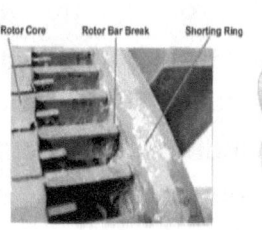

Industrial Electric Motors: By Mohammed H.A. Soliman

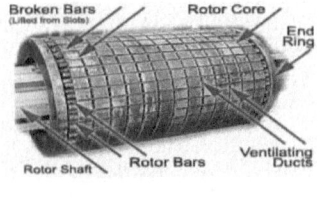

Defect rotor bars

A. Rotor Defects.
 Broken rotor bars
 Eccentric rotor

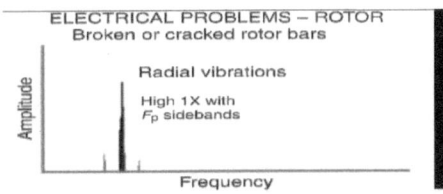

High 1X with FP sidebands

Broken rotor bars

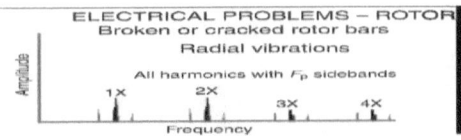

Broken rotor bars
All harmonics with FP sidebands

Industrial Electric Motors: By Mohammed H.A. Soliman

Example:

Motor Speed Synchronous = 1800RPM
Motor Speed Actual = 1770RPM
No of poles=4
FL=60hz
FP= 2xFL/P-RPM*P=2hz

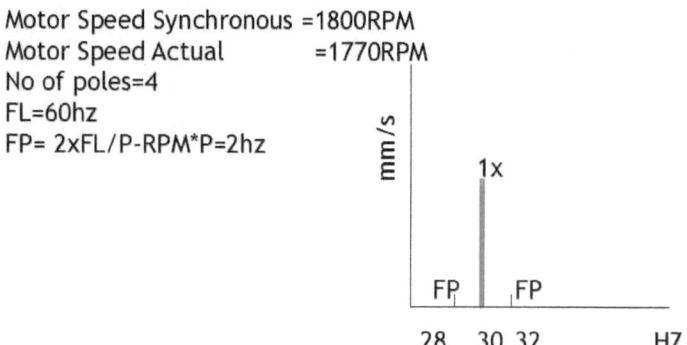

B. Eccentric rotor

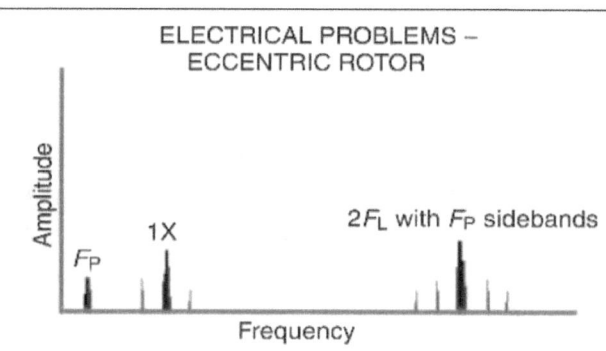

Industrial Electric Motors: By Mohammed H.A. Soliman

Component Name & Function : Insulation & winding, carry current

Failure Mode	Failure Effect	Severity	Failure cause	Failure Cause	Failure Cause	Occurrence	Current Controls	Detection	RPN
Winding failure/shortage	Motor failure	8	Overheat	Bad maint	Wear	2	Basic measurements include voltage	6	96
			Moisture	Bad maint					
			Contamination						
			Insulation breakdown						
			High vibration	Bearing	Bad maint	3		10	240
				Misalignment		2		10	160
			Voltage surges	aging	Bad maint	1		10	80

Recommendation	Take actions	Result			
		S	O	D	RPN
Monitor machine vibration on regular basis	Use vibration analysis to monitor bearing condition and shaft misalignment	8	3	1	24
		8	2	1	16

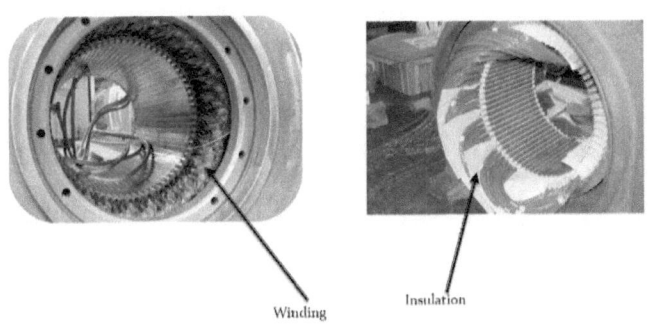

Winding   Insulation

Industrial Electric Motors: By Mohammed H.A. Soliman

Component Name & Function: Fan, keep the motor temp down in order for the motor components to perform well

Failure Mode	Failure Effect	Severity	Failure cause	Sources of failure	Failure Cause	Occurrence	Current Controls	Detection	RPN
Fan failure	Overheating and lead to expensive repair	8	Corrosion	Environmental issue	Aging	1	Visual inspection & cleaning	2	16
			Physical damage	Crash		1		2	16
				Carless handling					
			Foreign material build up	Lack of cleaning	Bad maintenance	1		2	16
						1		2	16

Industrial Electric Motors: By Mohammed H.A. Soliman

RPN Analysis for Electric Motor Failure Components

Part/Item	RPN
Bearing	64
	80
	320
	8
	160
	128
Shaft & Coupling	48
	80
	64
	80
Stator	400
	400

Part/Item	RPN
Rotor	80
	80
Winding & Insulation	96
	240
	160
	80
Cooling Fan	16
	16
	16
	16
Total	**2624**

A cutoff point of RPN 160 can be set because this will achieve over 50% improvement to risk number.

Industrial Electric Motors: By Mohammed H.A. Soliman

A cutoff point of RPN 160 can be set because this will achieve over 50% improvement to risk number.

Expected Total Risk Priority Number after applying the corrective actions:

RPN Reduction % = $R_{initial} - R_{revised} / R_{initial}$
$= 2624 - 1040 / 2624$
$= 60\%$

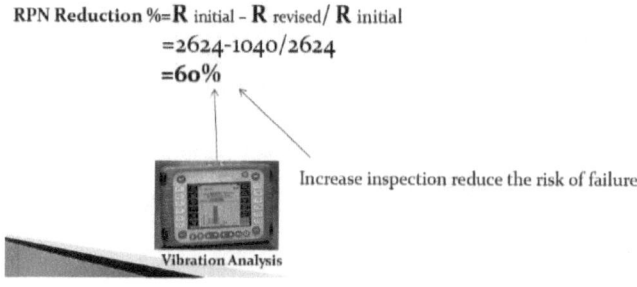

Increase inspection reduce the risk of failure

Vibration Analysis

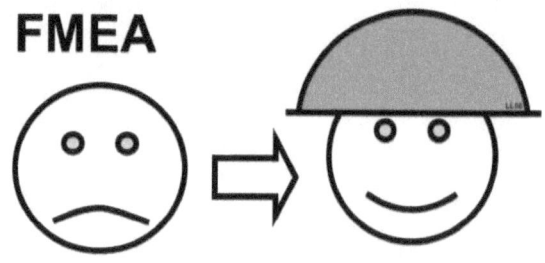

The improvements that yielded success included using vibration analysis to detect electrical issues and using infrared analysis to detect mechanical damages.

Industrial Electric Motors: By Mohammed H.A. Soliman

Electric Motor Predictive Maintenance

You might have seen in the previous FMEA case study, how predictive maintenance was utilized to improve the reliability of the electric motor. Increasing detection at early stages, also improving inspection and maintenance, increase the reliability of equipment or machine.

Infrared Thermography
Infrared monitoring and analysis have the widest range of application (from high- to low-speed equipment), and it can be effective for spotting both mechanical and electrical failures. It also requires minimum skills for analysis.

Applying the technique for power transformers involve using a thermography camera to inspect certain parts like: insulation, bushings, windings, connectors.

How to start?

- ✓ Make reference for the temp of all equipment when installing them in a new plant by taking photos and recording temperatures.
- ✓ If the above step were skipped or haven't been performed, ask the manufacturer for the temp reference of all equipment parts under normal operation.
- ✓ If the above tip is not possible, use the comparison method by comparing a transformer with the one beside it, same for any other equipment fans, compressors, pumps…etc. Use the standby maintained equipment as a reference.

Industrial Electric Motors: By Mohammed H.A. Soliman

Thermography Elements to Inspect	Failure Type
Couplings	Worn
Belt Conveyors	Loose or over tightened
Motor Bearing	Overheating, low grease, lubrication issues, failure
Roller Bearing	Overheating, law grease, bad grease

Industrial Electric Motors: By Mohammed H.A. Soliman

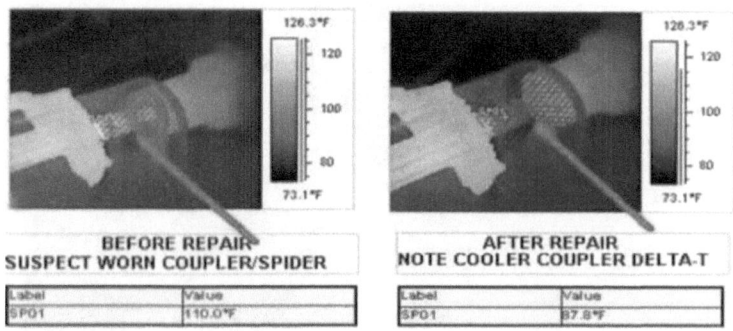

Temp analysis indicates worn coupling

Transmission belts

Motor belt is overheated, possible reasons are loose belt or over tightened

Infrared thermography facilitate inspection, it's really a fast and reliable method for inspecting thousands of bearings and equipment every day.

In electric motors, the consequence of failure for motor bearings can be huge at the end. An electric motor that is used to produce ore manufacture a product. Breakdowns can delay productivity, therefore delay the product to customer result in unhappy customer or losing few customers!

Pump Bearing

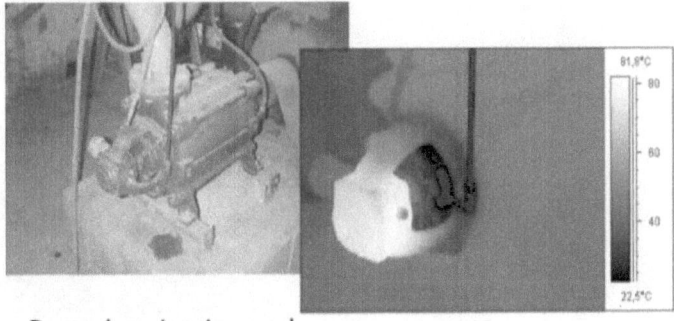

- Pump bearing in need of maintenance.

Electric Motor Bearing

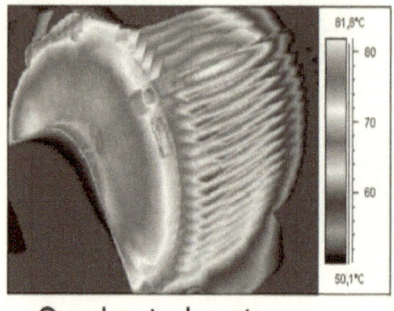
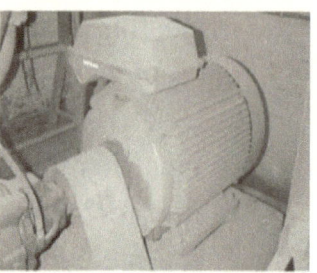

- Overheated motor bearing. Over 80 °C on bearing housing.

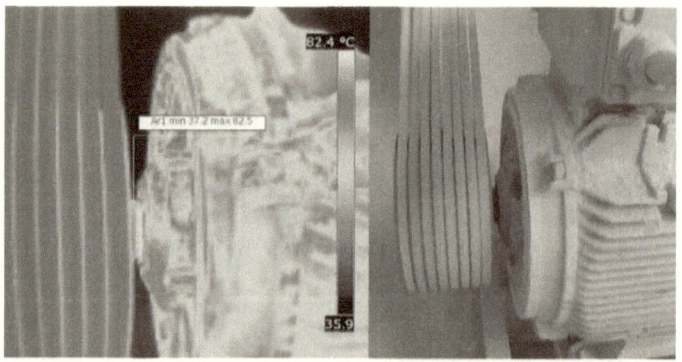

Motor bearing temperature is high due to excessive belt tension

Industrial Electric Motors: By Mohammed H.A. Soliman

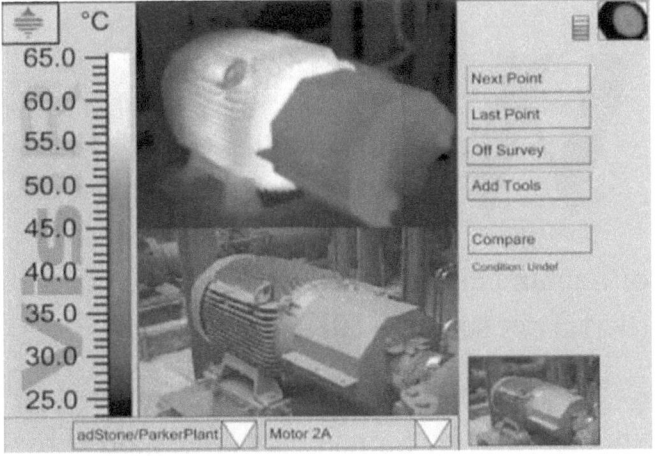

Ultrasound Analysis

What is Ultrasound?

Ultrasound is cyclic sound pressure with a frequency greater than the upper limit of human hearing, excess of 20,000 cycles (hertz) per second (20KHZ).

So, by definition, ultrasound is totally undetectable by human ears unless aided by instruments capable of translating ultrasound to audible sound. In the marketplace, these instruments are commonly known as ultrasonic detectors and have been used for various maintenance related functions for over 25 years.

Ultrasonic is a predictive maintenance technique and one of the non-destructive testing tools that used in the field of industry to detect early & hidden equipment failures.

Ultrasound Detector
Lightweight and portable, ultrasonic

translators are often used to inspect a wide variety of equipment. Some helpful accessories are supplied with the instrument too.

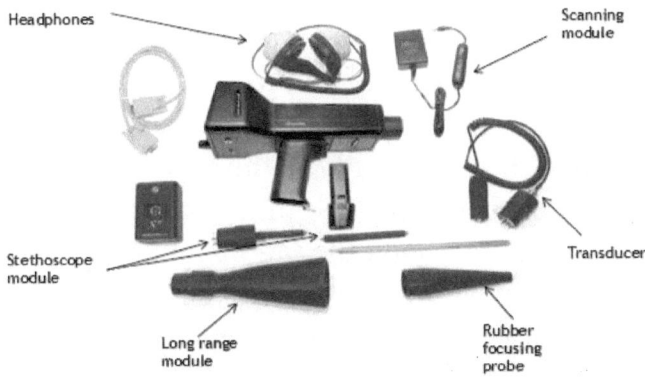

This is called Ultra probe, available in different type for wide range of uses.

Industrial Electric Motors: By Mohammed H.A. Soliman

Ultrasonic Condition-Based Lubrication

Ultrasound innovation is ideally appropriate for condition-based oil techniques. With ultrasonic review instruments a program can be set up that will educate monitors which bearing should be greased up and help oil professionals realize precisely how much oil to apply.

To see how these instruments can function successfully in the loud environments of a run of the mill plant, one must comprehend the innovation of ultrasound, how ultrasound is produced by direction, and how ultrasound observing instruments can help keep up ideal oil levels in bearing.

The innovation depends on the detecting of high-recurrence sounds. Ultrasound is considered to begin at 20,000 cycles for each second, or 20 kilohertz (kHz). This is viewed as the high-recurrence edge at which human hearing stops. Most ultrasonic instruments utilized to screen hardware will detect from 20 kHz up to 100 kHz. The scope of human hearing spreads frequencies of from 20 cycles for every second (20 Hz) up to 20 kHz. The normal human will frequently hear up to 16.5 kHz and no more.

These recurrence correlations are critical to note on the grounds that there are contrasts in the manner low-recurrence and high-recurrence sounds travel, which assist us with understanding why ultrasound can be successfully established in bearing checking and oil programs.

Oiling Procedures

It is basic to think about two components of expected failure: absence of oil and over grease. Ordinary bearing burdens cause a versatile disfigurement of the components in the contact region giving a smooth circular circulation. Yet, bearing surfaces are not entirely smooth. Consequently, the genuine pressure appropriation in the contact zone will be influenced by an arbitrary surface harshness. Within the sight of a grease film on a heading surface, there is a hosing impact on the pressure conveyance, and the acoustic vitality delivered will be low. Should oil be diminished to a point where the pressure conveyance is not, at this point present, the typical unpleasant spots will connect with the face surfaces and increment the acoustic vitality. These ordinary infinitesimal distortions will start to deliver wear and the conceivable outcomes of little gaps may create which adds to the "pre-failure" condition. Accordingly, beside typical wear, the weakness or administration life of an orientation is firmly affected by the relative film thickness gave by a fitting ointment.

Staying away from Over Lubrication
At the point when an excessive amount of ointment is placed into the bearing lodging, pressure constructs up and can prompt an expansion of warmth, which can make pressure and disfigurement of the bearing. Or then again it can break or "pop" the bearing seal permitting grease to spill out into undesirable zones, (for example, an engine winding), or permit foreign substances to enter the raceway. All of which can prompt bearing failure.

The suitable measure of oil is significant. On the off chance that a bearing is over greased up the bearing can be pushed unnecessarily by the ointment causing extra wear of the bearing. Then again, if there isn't sufficient grease, the bearing will rub on the strong surface, again causing erosion and wear on the orientation. Either case is hindering to the life of the bearing. Utilizing airborne/structure borne ultrasound removes the speculation from grease.

Ultrasound Monitoring

Ultrasound instruments identify changes identified with contact. An appropriately greased up bearing will have next to no grating. The ointment levels out any pressure the bearing experiences as it moves around the raceway in this manner decreasing the potential for ruinous contact. As the bearing moves, it delivers a conspicuous "surging" sound much the same as the sound of air spilling out of a tire. This surging sound is alluded to as "background noise." incorporates all sounds, both low and high frequencies. The high-recurrence waves created by this background noise more restricted than those of the lower frequencies. Utilizing a ultrasonic interpreter, these signs can be identified with practically zero impedance from other mechanical commotions created by different parts, for example, a pole or another bearing close by. As the oil level in a heading falls or weakens, the potential for contact increments. There will be a relating ascend in the ultrasound adequacy level that can be noted and heard. The technique to decide when to grease up.

Furthermore, when to quit applying grease with ultrasound instruments is as straightforward as: setting a gauge, setting assessment timetables and checking as you grease up.

Setting up the greasing level:
A benchmark for an orientation reflects in decibels the level at which it is working under typical conditions with no detectable faults and with sufficient grease.

There are three techniques for setting a pattern

1. Examination: when there is more than one orientation of a similar sort, load what's more, rpm, various orientation can be contrasted one with the other. Each bearing is examined at a similar test point and edge. The decibel levels furthermore, stable quality is looked at. On the off chance that there are no considerable contrasts, (under 8dB) a pattern dB level is set for each bearing. This is normally performed with a compact ultrasonic interpreter.

2. Set while greasing up. While oil is being applied, tune in until the sound level drops down and starts to rise. By then no more oil is included and the dB esteem is utilized as the pattern.

3. Historical: bearing dB levels are gotten from an underlying review. After thirty days the bearing dB levels are taken and looked at. On the off chance that there is close to nothing (under 8dB) to no adjustment in dB than the benchmark levels are set and will be utilized for examination for resulting assessments.

Setting Inspection Schedules
This should be based on the equipment criticality, environment, type of industry, failure consequence, failure occurrence and the availability of standby. Typically, one month is good. But for baring that have had significant levels and have been along these lines greased up, it may be important to test all the more habitually to take note of any potential changes.

Accessibility Problems

There might be circumstances in which it might be hard to access a few bearings. For instance, there might be a perplexing machine where a bearing is inserted in a territory where just a lube tube is reached out external the packaging. On the off chance that the lube tube is a conductive metal, for example, copper, the bearing can even now be tried and a grease activity level set. On the off chance that the fitting is of a non-sound conductive material, for example, plastic, a different conductive metallic wave guide can be introduced so the bearing can be observed. The wave guide can be confined from structure borne clamor of the machine (the mounting point) through elastic disconnection material. Should it not be conceivable to put a wave manage, there is an elective arrangement. A transducer can be forever mounted on the bearing lodging and a link race to an opening. The link can be appended to a specific connector that can be "connected" to the ultrasonic sensor, as demonstrated as follows.

Auto Greaser and Ultrasound Sensor

Some systems prefer to use auto greasers for bearing. In this case installing an ultrasound sensor is a must to adjust the grease flow. The grease will be on and off based on the ultrasound sensor orders.

Conclusion

Ultrasound innovation is perfectly appropriate for viable condition-based grease programs. The short-wave nature of the sign diminishes obstruction from contending commotions and permits reviewers to precisely screen bearing condition. By setting up a caution level of 8 dB over a given benchmark, investigators will know when and when not to grease up. Over oil can be dodged by applying just enough grease to accomplish standard levels or tune in to a drop in the sound level should no dB reference be accessible.

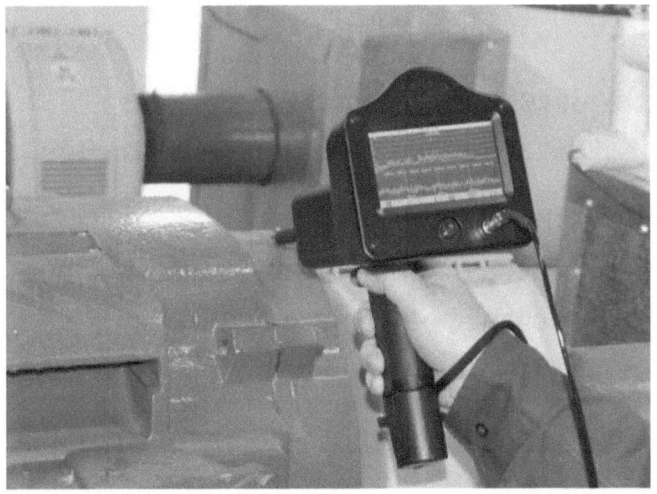

Vibration Analysis

Vibration can be defined as simply the cyclic or oscillating motion of a machine or machine component from its position of rest.

Displacement:
It is the amplitude of a point on a structure.

Velocity:
Is the speed of a point in a system, It is the rate of change of displacement.

Acceleration:
Is the rate of change of velocity of a point in a system.

Frequency:
Is the no of cycles (vibrations) per second, measured by hertz (HZ).

Harmonics:
Frequency component at a frequency that is an integer (whole number e.g., 2X. 3X. 4X,

etc.) multiple of the fundamental (reference) frequency.

How many RPMs in 1 hertz?
Since hertz is in sec, and RPM is in minute, 1 Hz= 60 RPM.

How to convert RPM to Hertz?
1500 RPM = 25 Hz.

Why we measure the vibration?
- ✓ To detect what is out of the human sense.
- ✓ To discover hidden failures.
- ✓ To Detect early failures & monitor the machine health condition.
- ✓ To assure the quality of repairs.
- ✓ As a useful tool to improve the maintenance reliability.

Common Industrial Applications:
- Pumps.
- Fans.
- Turbines.
- Agitators.
- Stirrers.
- Compressors.
- Electric. Motors.
- Gearboxes.

Vibration& Reliability

Vibration analysis alone doesn't improve reliability, root cause analysis and acceptance testing can help.

There are two ways that we can utilize vibration to improve reliability:

First, if we study the vibration, we can often determine why the fault condition developed in the first place; for example, what caused the crack to appear in the inner race of the bearing? If we perform root cause failure analysis, we can make chances to our proactive so that the bearing doesn't suffer the same fate in the future.

Second, when we overhaul the machine, we can again use the vibration analysis to check that the maintenance repair has been made correctly; and that the machine is correctly aligned and balanced, this called acceptance testing.

Vibration is still used to monitor the health of the machine, but if we improved the reliability of the machine, we will see fewer faults' conditions develop.

Measuring With Smart Sensors "Collecting Data"

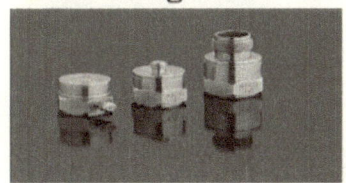 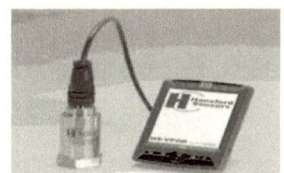

Analysis with Smart Software

Industrial Electric Motors: By Mohammed H.A. Soliman

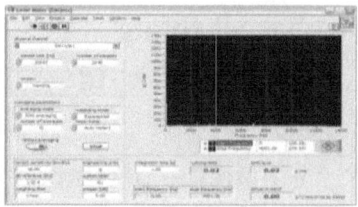 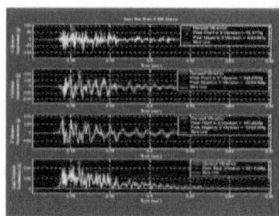

Getting Results

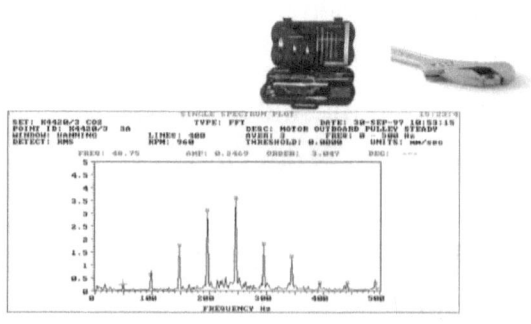

Performing Actions accordingly
Maintain
Repair
Cure

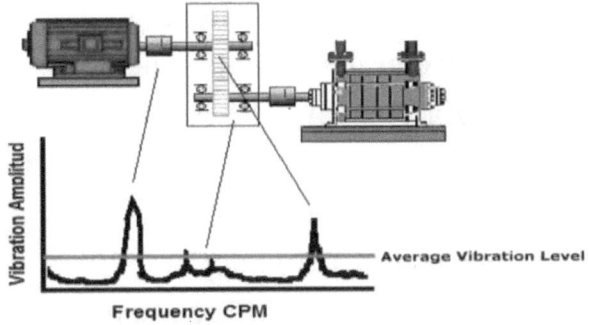

What to Measure??
We normally measure the vibration speeds in in/sec or mm/s.

Why using Condition Monitoring programs (predictive maintenance)? And why particularly Vibration Analysis technique?

Benefits of setting up a Predictive Maintenance (Pd.M.) program:

1. To detect what is out of the human sense.
2. To discover hidden failures.
3. To Detect early failures & monitor the machine health condition.
4. To reduce Maintenance Costs.
5. As a useful tool to improve the machine reliability.

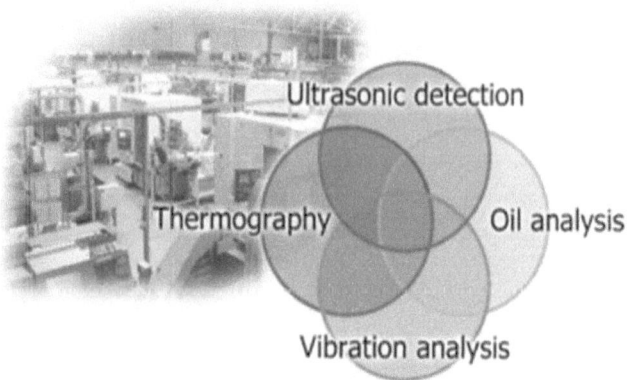

Four tools make up 85% of any PdM program.

Industrial Electric Motors: By Mohammed H.A. Soliman

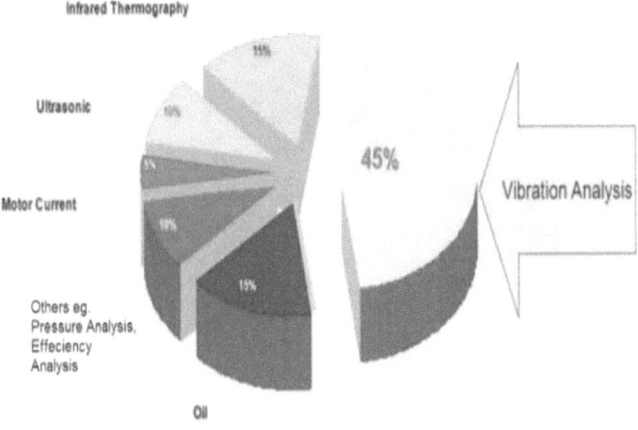

Vibration presents 45% of PdM programs.

Equipment that fails in service can cost up to 10 times more to repair than the equipment repaired when predicted by condition monitoring.

Other PdM Techniques

Notes:

Motor diagnosis = motor current analysis, and it's a technique involve intensive diagnosis of motor currents.

Oil Analysis involve Wear Particles Analysis for more intensive diagnosis about the sources of failure. For more information about the technique read the book: Machinery Oil Analysis and Condition Monitoring.

Thermography: involve thermal analysis using infrared camera. For more information about the technique, read the book: Industrial Applications of Infrared Thermography.

Ultrasound Analysis: is an acoustic method based on high frequencies measurement. For more information, read the book: Ultrasound Analysis for Condition Monitoring.

Industrial Electric Motors: By Mohammed H.A. Soliman

Why Vibration?

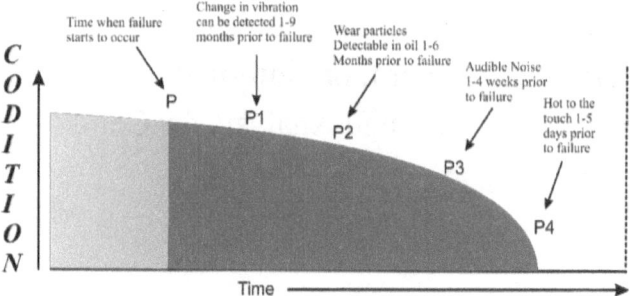

Vibration VS Thermography VS Oil Analysis

Type of fault	Vibration	Temp	Oil
Out of balance	xxx	----	----
Misalignment	xxx	x	----
Damage of bearing	xxx	xx	x
Damage of gear box	xxx	x	xx
Belt problems	xx	----	----
Motor problems	xx	x	----
Mechanical looseness	xxx	x	x
Resonance	xxx	----	----

Industrial Electric Motors: By Mohammed H.A. Soliman

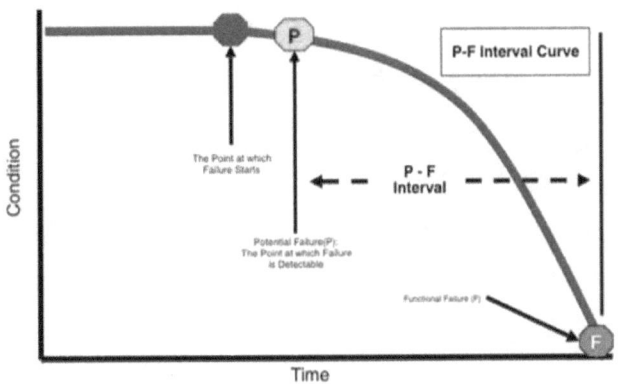

One of the most benefits of a condition monitoring program is to detect potential failures at early state.

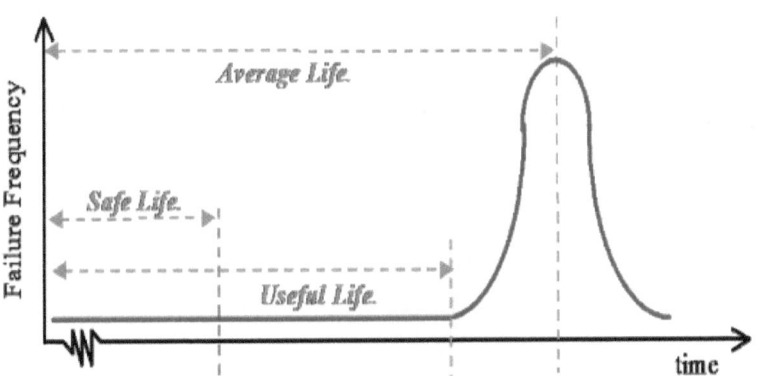

Determine the PM Interval Using Reliability Data from PdM Programs.

On Site Tools Used for Measurements & Analysis of the Mechanical Vibration

Simple tools, Vibration pen
- Simple.
- Accurate.
- Easy.
- Less Expensive.
- Quick Measure.

Simple but with more data and info
- ✓ Quick spectrum view
- ✓ Compare vibration reading to the charts that indicate the severity of the faults according to the type and size of the machine.
- ✓ Inexpensive way to start with vibration measurement.

Industrial Electric Motors: By Mohammed H.A. Soliman

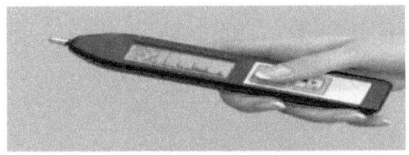

Machines Vibration Limits and Reference

Class I (Small): Machines less than 20HP.
Class II (Medium): Machines from 20-100HP without special foundations.
Class III (Large): Machines with rigid foundations and over 100HP.
Class IV (Large): Machines with soft foundation and over 100HP.

Machine		Class I small machines	Class II medium machines	Class III large rigid foundation	Class IV large soft foundation
in/s	mm/s				
0.01	0.28				
0.02	0.45				
0.03	0.71		good		
0.04	1.12				
0.07	1.80				
0.11	2.80		satisfactory		
0.18	4.50				
0.28	7.10		unsatisfactory		
0.44	11.2				
0.70	18.0				
0.71	28.0		unacceptable		
1.10	45.0				

Vibration Velocity Vrms

VIBRATION SEVERITY PER ISO 10816

Professional Detectors/Analyzers

- ✓ Professional.
- ✓ More Accurate.
- ✓ Software Analysis.
- ✓ Automatic Analysis.
- ✓ Comprehensive results.
- ✓ Costly.

Please discuss with the vendor the suitable instrument for your business!

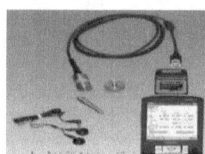

Industrial Electric Motors: By Mohammed H.A. Soliman

Portable professional analyzers with ERP connected.

Dual Channel Analyzer.
Single Channel Analyzer.
USB Connectivity.

Measurement Techniques (points of measuring)

Measuring at bearing points in three directions:
Horizontal.
Vertical.
Axial.

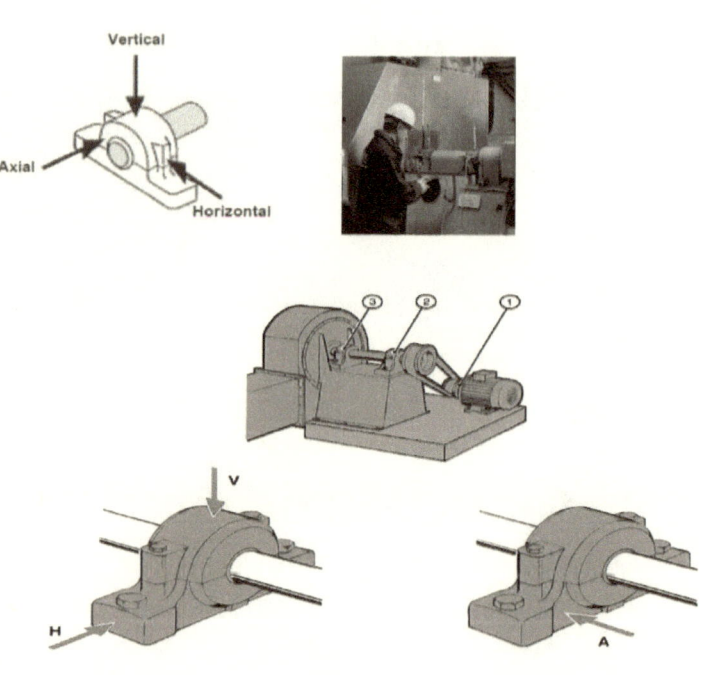

Why measure in 3 directions?

Taking readings in three directions gives more information that helps in analysis as some defects comes with predominant vibrations in a particular direction.

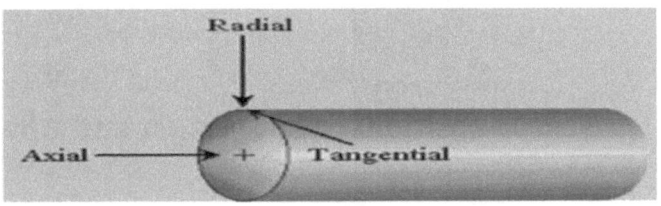

Axial is the direction parallel to the centerline of a shaft or turning axis of a rotating part. Radial is that direction toward the center of rotation of a shaft or rotor. The Tangential measurement is that measurement that is tangent or perpendicular to the radial transducer.

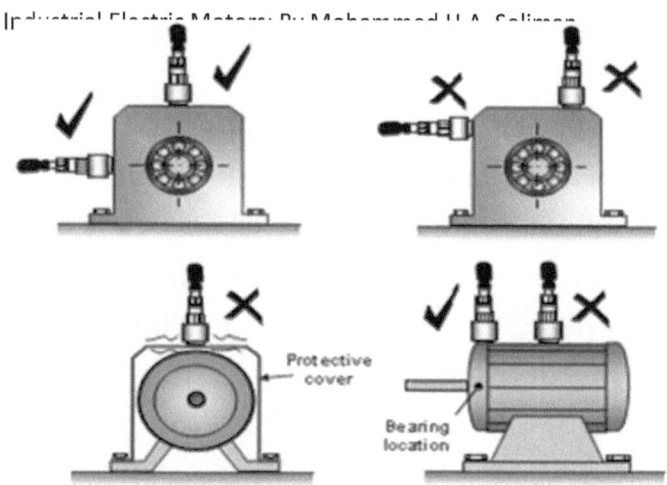

Sensor positions, what is right and what is wrong

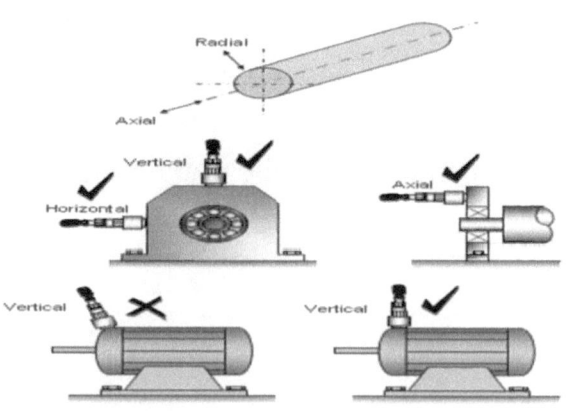

Vibration Standards

Class I (Small): Machines less than 20HP.
Class II (Medium): Machines from 20-100HP without special foundations.
Class III (Large): Machines with rigid foundations and over 100HP.
Class IV (Large): Machines with soft foundation and over 100HP.

Vibration Reports & Standards

Standard acc to ISO 10816

Machine		Class I small machines	Class II medium machines	Class III large rigid foundation	Class IV large soft foundation
in/s	mm/s				
0.01	0.28				
0.02	0.45				
0.03	0.71	good			
0.04	1.12				
0.07	1.80				
0.11	2.80	satisfactory			
0.18	4.50				
0.28	7.10	unsatisfactory			
0.44	11.2				
0.70	18.0				
0.71	28.0	unacceptable			
1.10	45.0				

(Vibration Velocity Vrms)

Industrial Electric Motors: By Mohammed H.A. Soliman

Quick Example – Centrifugal Fan

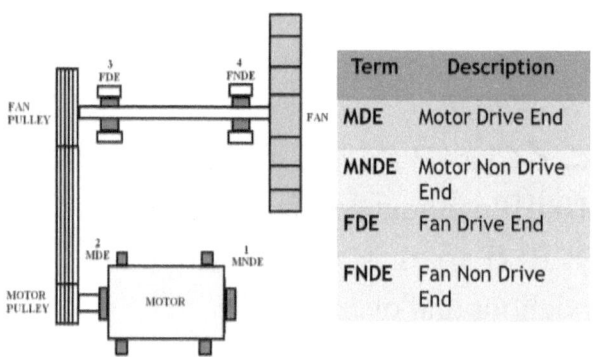

Term	Description
MDE	Motor Drive End
MNDE	Motor Non Drive End
FDE	Fan Drive End
FNDE	Fan Non Drive End

Ex. General Components of centrifugal fan

Vibration Measurements

POINT	DESCRIPTION	overall values	LIMITS
1 H	E-Motor non drive end horizontal	30.468 mm/s	x x x
1 V	E-Motor non drive end vertical	13.611 mm/s	x x x
1 A	E-Motor non drive end axial	-	
2 H	E-Motor drive end horizontal	24.302 mm/s	x x x
2 V	E-Motor drive end vertical	23.217 mm/s	x x x
2 A	E-Motor drive end axial	22.827 mm/s	x x x
3 H	fan fixed bearing coupling side horizontal	35.610 mm/s	x x x
3 V	fan fixed bearing coupling side vertical	31.521 mm/s	x x x
3 A	fan fixed bearing coupling side axial	-	
4 H	Fan free bearing fan side horizontal	29.609 mm/s	x x x
4 V	Fan free bearing fan side vertical	26.941 mm/s	x x x
4 A	Fan free bearing fan side axial	24.733 mm/s	x x x

DANGEROUS	x x x
ALARM	x x
ACCEPTED	x

Spectrum chart

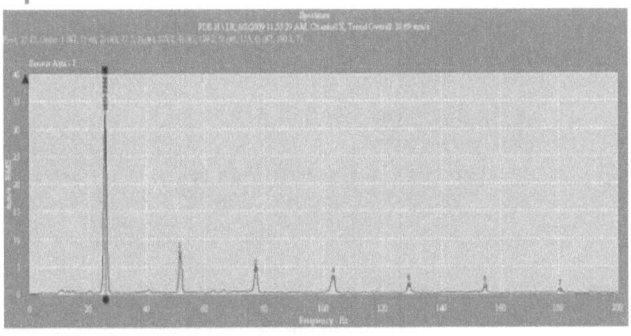

Fig.
Machine showing significant increase in vibration level at frequency of 25hz indicating unbalance issue.

Action report/recommendations examples:

Perform Balancing.
Need to Replace Bearing no 1, 2...etc.
Check for Bearing Grease & lubrication condition.
Monitor Bearing Condition for Early Failure.
Perform Alignment (for shaft, pulleys...etc.).
Check for Foundation Looseness, internal looseness...etc.

Example.2 Discuss the Following Vibration Analysis Data Report

Equipment: Centrifugal Fan
Class: IV
Measurements are in mm/s

Parameters		1-9-2010	1-10-2010	1-11-2010	1-12-2010	1-1-2011
Point 1 (fan drive end)	x	1.411	2.835	4.7	9.5	12.7
	y	1.865	2.80	4.2	8.0	14.3
	z	2.487	4.853	4.9	7.3	7.2
Point 2 (fan non drive end)	x	1.490	3.0	4.87	6.5	14.0
	y	0.9	1.0	2.5	4.2	7.0
	z	1.854	2.2	4.1	7.1	12.0

System Description
Technical Data:
Fan Type=   Centrifugal fan
Flow rate=   78,000m3/h
Motor Power=   200KW
Horse power=   268HP
Motor Speed=   1500RPM
Fan Speed=   1500RPM
Transmission type: V-belts
Spare Parts Data:
FDE Bearing: 22222EK +H322

Industrial Electric Motors: By Mohammed H.A. Soliman

FNDE Bearing:	22222EK +H322
MDE Bearing:	6322
MNDE Bearing:	6322
Bearing Housing:	SNH 522-619
Pulleys type:	SPC335
V-belts type:	SPC5300

Equipment: Centrifugal Fan
Class: IV
Measurements are in mm/s

Parameters		1-9-2010	1-10-2010	1-11-2010	1-12-2010	1-1-2011
Point 1 (fan drive end)	x	1.411	2.835	4.7	9.5	12.7
	y	1.865	2.80	4.2	8.0	14.3
	z	2.487	4.853	4.9	7.3	7.2
Point 2 (fan non drive end)	x	1.490	3.0	4.87	6.5	14.0
	y	0.9	1.0	2.5	4.2	7.0
	z	1.854	2.2	4.1	7.1	12.0

Green color = Acceptable
Yellow color = Alarm
Red color = Dangerous

Remarks:

First month: Over all machine health is good.

Second month: Over all machine health is within the acceptable limits "but recommend to monitor bearing condition"

Third month: There is a significant increase in vibration speed, machine will require balancing.

Forth month: machine is in critical level and required immediate shutdown to perform the following:

   -Balancing.

   -checking for pulleys alignment and belt tensioning.

   -Monitor non-drive end bearing and drive end bearing for replacing. requirement or greasing.

Fifth month: Stop the machine to avoid sudden failure caused by one of the following issues: -

   -Crack in foundation.

   -Bearing failure.

-Housing wear.

Question: Why it's important to have all the machine data & spare parts information before processing the vibration analysis?

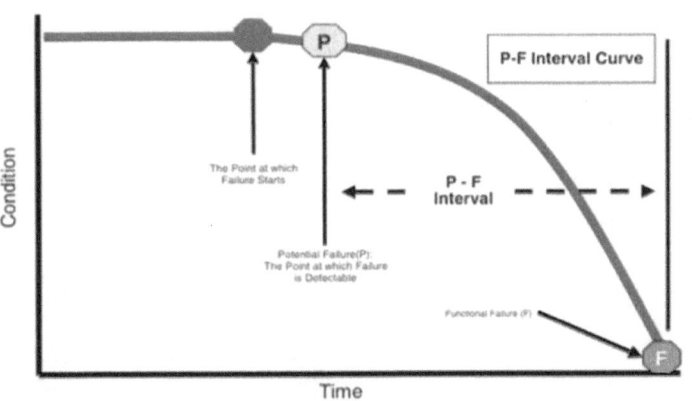

One of the most benefits of a Condition Monitoring program is to detect potential failures at early state.

Methodology of Measuring "Collecting Data"

By using a sensor called accelerometer & an electronic meter that record the vibration.

The accelerometer converts the physical vibration into an electrical signal that can be measured by meter, the meter can be very simple recording device or can be a data collector and analyzer that enable the person collecting the data to perform a wide range of tests whilst in the field. The meter and the sensor are taken out into the

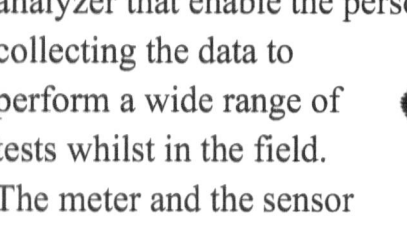

plant and the sensor is temporary mounted on each bearing (with the aid of strong magnet) and the vibration reading is taken.

A Simple meter display the vibration reading for comparison against alarms, but more advanced meters record the vibration for later analysis.

The Different Types of Vibration Sensors

1 -Velocity pickup
The velocity pickup is a very common transducer for monitoring the vibration of rotating machinery. This type of vibration transducer installs easily on most analyzers, and is rather inexpensive compared to other sensors. For these reasons, the velocity transducer is ideal for machine-monitoring. Velocity pickups have been used as vibration transducers on rotating machines.

For a very long time, and these are still utilized for a variety of applications today. Velocity pickups are available in many different physical configurations and output sensitivities.

Theory of Operation
When a coil of wire is moved through a magnetic field (coil-in-magnet type) a voltage is induced across the end wires of the coil. The transfer of energy from the flux

field of the magnet to the wire coil generates the induced voltage. As the coil is forced through the magnetic field by vibratory motion, a voltage signal correlating with the vibration is produced.

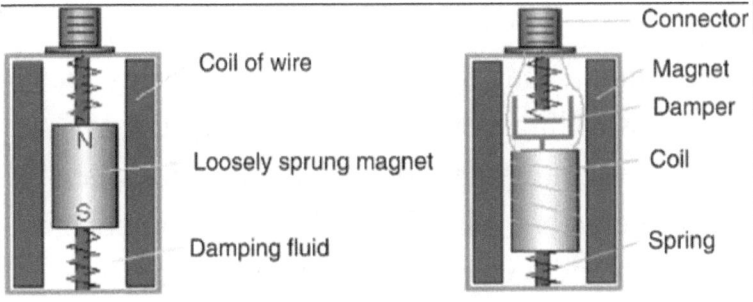

The magnet-in-coil type of sensor is made up of three components: a permanent magnet, a coil of wire and spring supports for the magnet. The pickup is filled with oil to dampen the spring action. The relative motion between the magnet and coil caused by the vibration motion induces a voltage signal.

Industrial Electric Motors: By Mohammed H.A. Soliman

2-Acceleration transducers/pickup

Accelerometers are the most popular transducers used for rotating machinery applications. They are rugged, compact, lightweight transducers with a wide frequency response range. Accelerometers are extensively used in many condition-monitoring applications. Components such as rolling element bearings or gear sets generate high vibration frequencies when defective. Machines with these components should be monitored with accelerometers. The installation of an accelerometer must carefully be considered for an accurate and reliable measurement.

ACCELEROMETER: Transducer whose output is directly proportional to acceleration. Most commonly used are mass loaded piezoelectric crystals to produce an output proportional to acceleration.

Theory of operation

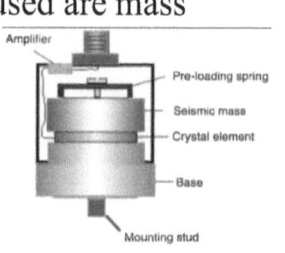

Accelerometers are inertial measurement devices that convert mechanical motion into a voltage signal. The signal is proportional to the vibration's acceleration using the piezoelectric principle. Inertial measurement devices measure motion relative to a mass. This follows Newton's third law of motion: body acting on another will result in an equal and opposite reaction on the first. Accelerometers consist of a piezoelectric crystal (made of ferroelectric materials like lead zirconate titanate and barium titanate) and a small mass normally enclosed in a protective metal case.

When the accelerometer is subjected to vibration, the mass exerts a varying force on the piezoelectric crystal, which is directly proportional to the vibratory acceleration. The charge produced by the piezoelectric crystal is proportional to the varying vibratory force. The charge output is measured in Pico-coulombs per g (Pc/g) where g is the gravitational acceleration. Some sensors have an internal charge amplifier, while others have an external charge amplifier. The charge amplifier converts the charged output of the crystal to a proportional voltage output in mV/g.

Crystals will generate measurable piezoelectricity when their static structure is deformed by about 0.1% of the original dimension.

Frequency range

Accelerometers are designed to measure vibration over a given frequency range. Once the particular frequency range of interest for a machine is known, an accelerometer can be ranges are also available selected. Typically, an accelerometer for measuring machine vibrations will have a frequency range from 1 or 2 Hz to 8 or 10 kHz. Accelerometers with higher-frequency.

Calibration for Both Types

Piezoelectric accelerometers cannot be recalibrated or adjusted. Unlike a velocity pickup, this transducer has no moving parts subject to fatigue. Therefore, the output sensitivity does not require periodic adjustments. However, high temperatures and shock can damage the internal components of an accelerometer.

Velocity pickups should be calibrated on an annual basis. The sensor should be removed from service for calibration verification. Verification is necessary because velocity pickups are the only industrial vibration sensors with internal moving parts that are subject to fatigue failure.

Vibration Sensors Connection Types

There are two types of connections, Wireless and Wired

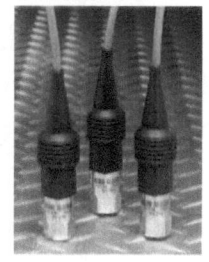

Wired Vibration Sensors Limitation:
- ✓ High cost of installation, especially in hazardous areas.
- ✓ Insufficient justification for a permanent system on certain balance-of-plant machines.
- ✓ Traversing/moving machines where fixed cabling is not possible.

Wireless Vibration Sensors Advantages:
- ✓ Overcome the wire sensor limitation.
- ✓ Facilitate applications that in the past were impractical, such as temporary

installations for troubleshooting and remote monitoring.
- ✓ Eliminates the need for hardwiring for communications.
- ✓ Reduced installation cost.

Challenges to Wireless Vibration Sensors:
- ✓ High bandwidth is needed, due to the relatively large amounts of data that need to be sent over the wireless link.
- ✓ Higher-level processing capabilities, and the ability to capture data at the right time are also key requirements.
- ✓ Battery-powered devices that are required to provide onboard power must satisfy customer demands for long service life.
- ✓ Wireless security is a must.

Tips:
The devices and sensors, as well as the wireless network components, must also

cope with conditions commonly found in the industrial environment, such as exposure to water, elevated temperatures, electrical interference, hazardous-area classifications, obstructions, physical location and distance.

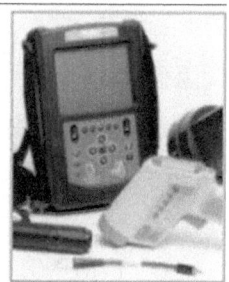

Advantages
- ✓ Can collect, record and display vibration data such as FFT spectra, overall trend plots and time domain waveforms.
- ✓ Provides orderly collection of data.
- ✓ Automatically reports measurements out of pre-established limit thresholds.
- ✓ Can perform field vibration analysis.

Disadvantages
- ✓ They are relatively expensive.
- ✓ Operator must be trained for use.
- ✓ Limited memory capability and thus data must be downloaded after collection.

Vibration analysis – database management software

The data collector/analyzer can collect and store only a limited amount of data. Therefore, the data must be downloaded to the computer to form a history and long-term machinery information database for comparison and trending. To perform the above tasks, and also aid in collection, management and analysis of machinery data, database management software packages are required.

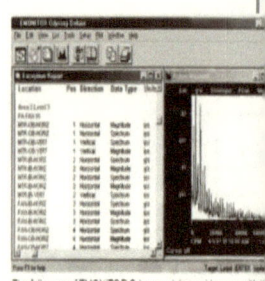

The full range of EMONITOR Odyssey plots provides you with the most complete tools for machinery analysis.

These database management programs for machinery maintenance store vibration data and make comparisons between current measurements, past measurements and predefined alarm limits. Measurements transferred to the vibration analysis software are rapidly investigated for deviations from normal conditions. Overall vibration levels, FFTs, time waveforms.

Reports can be generated showing machines whose vibration levels cross alarm thresholds. Current data are compared to baseline data for analysis and also trended to show vibration changes over a period of time. Trend plots give early warnings of possible defects and are used to schedule the best time to repair.

Advantages
- They aid in data collection, management and analysis of machinery data.
- They can save long-term machinery data that help to compare present and past condition-monitoring data.
- They assist in vibration analysis.
- They provide user-friendly reports.

Disadvantages
- The software programs are expensive, with sometimes almost the same cost as the data collector/analyzer hardware.

- They must be configured for individual requirements. A lot of information is required as initial input.

Online data acquisition and analysis (for Critical Machines)

Critical machines, as defined in the earlier topics, are almost always provided with continuous online monitoring systems. Here sensors (e.g., Eddy current probes installed in turbo-machinery) are permanently installed on the machines at suitable measurement positions and connected to the online data acquisition and analysis equipment. The vibration data are taken automatically for each position and the analysis can be displayed on local monitoring equipment, or can be transferred to a host computer installed with database management software.

This ability provides early detection of faults and supplies protective action on critical machinery.

Advantages
- ✓ Performs continuous, online monitoring of critical machinery.
- ✓ Measurements are taken automatically without human interference.
- ✓ Provides almost instantaneous detection of defects.

Disadvantages
- ✓ Reliability of online systems must be at the same level as the machines they monitor.
- ✓ Failure can prove to be very expensive.
- ✓ Installation and analysis require special skills.
- ✓ These are expensive systems.

Industrial Electric Motors: By Mohammed H.A. Soliman

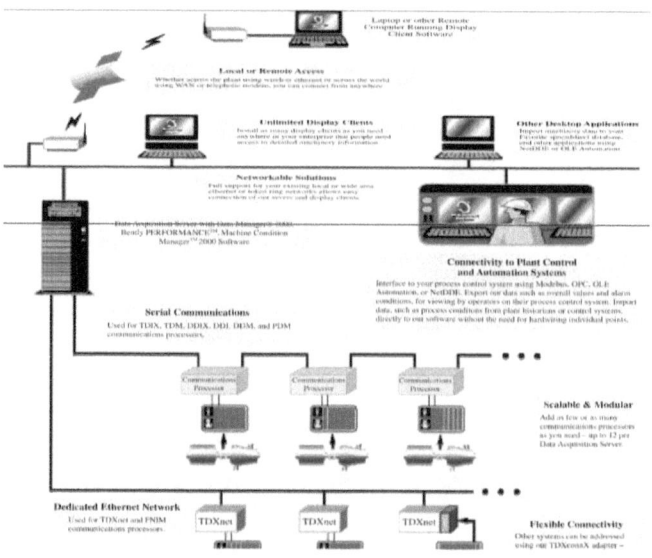

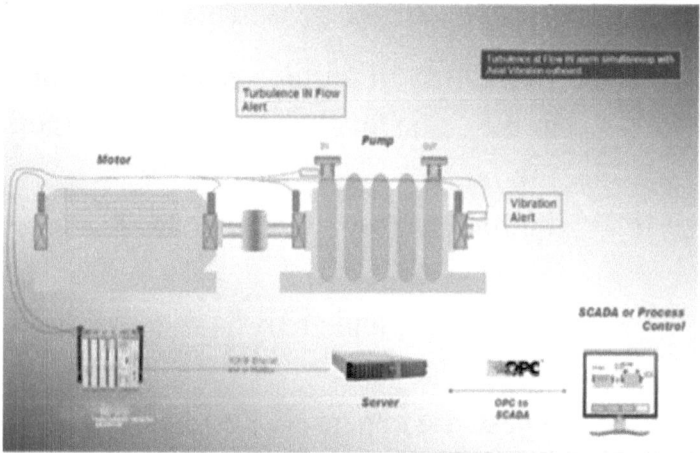

Vibration can be made online so you can have permanent monitoring to the critical machine's health & condition.

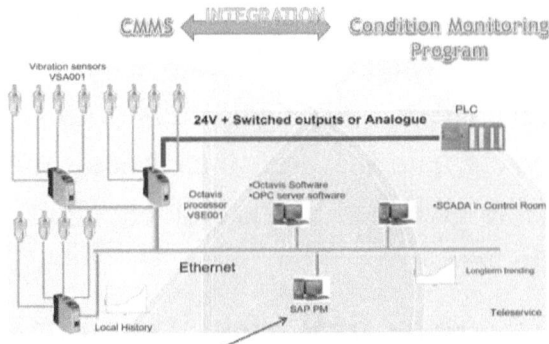

Automatic initiation of work orders depend on the machine condition

Vibration Resources
- Labor (technicians & engineers).
- Time Planning.
- Tools (Vibration Analyzer/ Vibration detector + Sensor + Software).
- Training.
- Cost Analysis.
- Budget planning & preparation.

Vibration Analysis-Signal Processing

The vibration of a machine is a physical motion. Vibration transducers convert this motion into an electrical signal. The electrical signal is then passed on to data collectors or analyzers. The analyzers then process this signal to give the FFTs and other parameters.

To achieve the final relevant output, the signal is processed with the following steps:
- Analog signal input.
- Anti-alias filter.
- A/D converter.
- Overlap.
- Windows.
- FFT.
- Averaging.
- Display/storage.

Time Waveform

A time waveform is the time domain signal. In vibration terms, it is a graph of displacement, velocity or acceleration with respect to time. The time span of such a signal is normally in the millisecond range. A graphical representation of a wave obtained by plotting some characteristic of the wave (such as amplitude), versus time.

The raw signal from the sensor is the time waveform

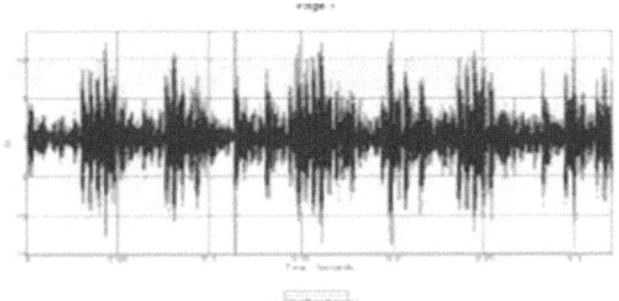

Industrial Electric Motors: By Mohammed H.A. Soliman

It's quite difficult to understand the time waveform. Time Waveform is an important analysis tool in certain circumstances.

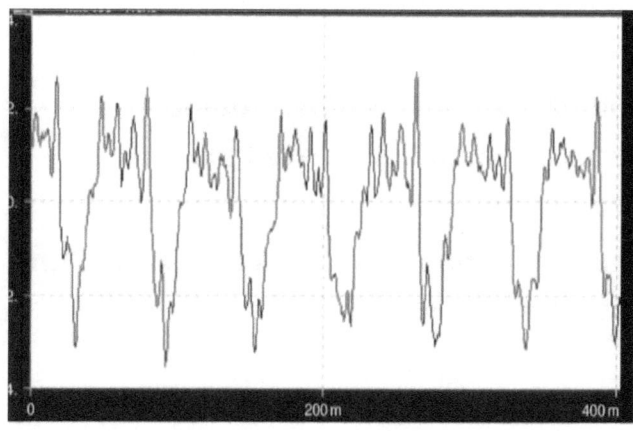

Enveloping and Demodulation Spectrum

This technique of vibration analysis is extensively used for fault detection in bearings and gearboxes. This method focuses on the high-frequency zone of the spectrum. Using a high-pass filter (allows high frequencies but blocks lower ones), the analyzer zooms into the low-level high-frequency data.

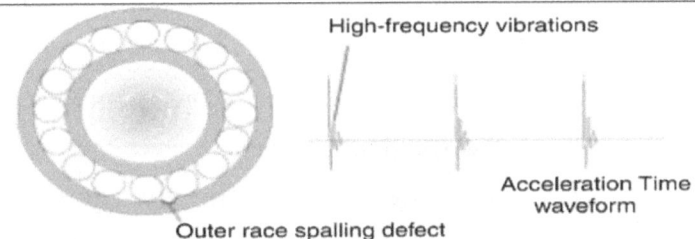

Industrial Electric Motors: By Mohammed H.A. Soliman

Phase Reading

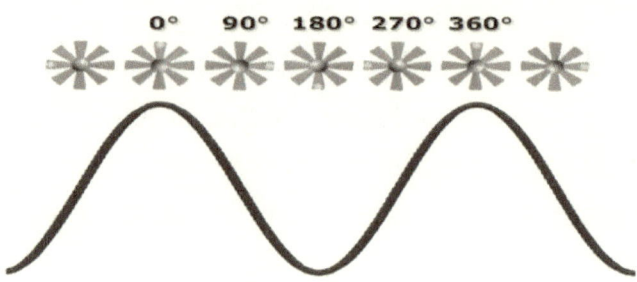

Frequency Spectrum

A spectrum is a graphical display of the frequencies at which a machine component is vibrating.

It's the FFT of the time waveform which produce the spectrum.

The data collector takes the raw time waveform from the sensor and perform a calculation called FFT (Fast Fourier Transform). This process extract all of the individual frequencies from the vibration pattern.

When machines run, it generates vibrations at different frequencies, the shaft turn is one frequency, the vans on the pump generate higher frequency, and the bearing generate another frequency.

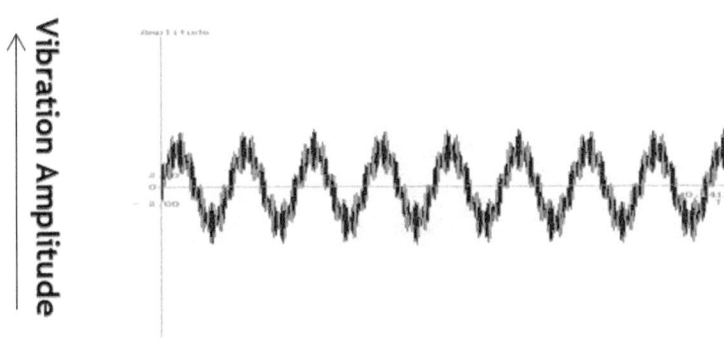

Industrial Electric Motors: By Mohammed H.A. Soliman

Signal Processing Flow

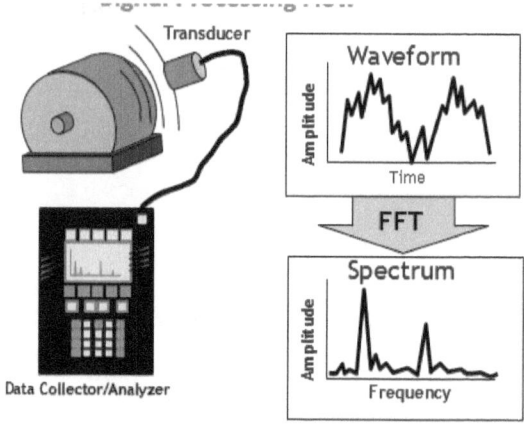

What is FFT?

The Fast Fourier Transform is a mathematical method for transforming a function of time into a function of frequency. Sometimes it is described as transforming from the time domain to the frequency domain. It is very useful for analysis of time-dependent phenomena.

Industrial Electric Motors: By Mohammed H.A. Soliman

Spectrum Analysis & Faults Diagnosis

If vibration amplitude turns to be increased at the bearing frequency, then we can determine what is wrong with the machine.

Machine Vibration Sources

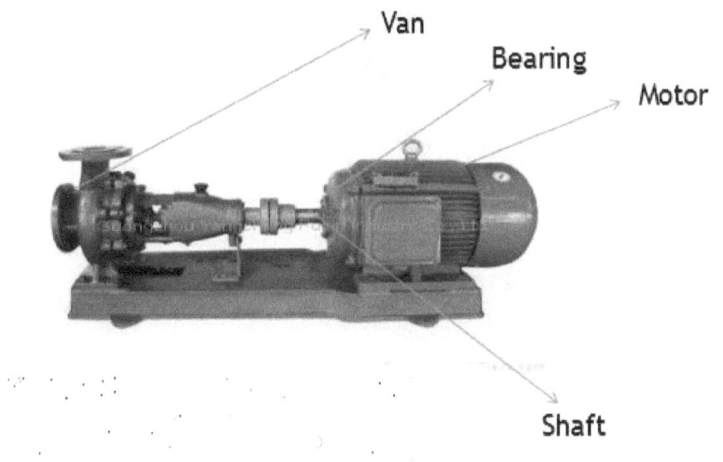

- Belt frequency 685 cpm
- Motor, 1× rpm 1485 cpm
- Fan shaft, 1× rpm 1000 cpm
- Motor, 4× rpm or electrical? 5964 cpm.

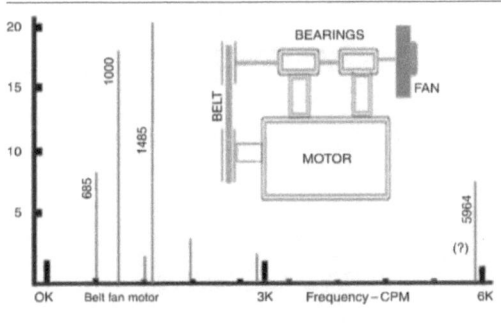

Industrial Electric Motors: By Mohammed H.A. Soliman

What are the Faults that Spectrum Analysis Can Tell Us About?

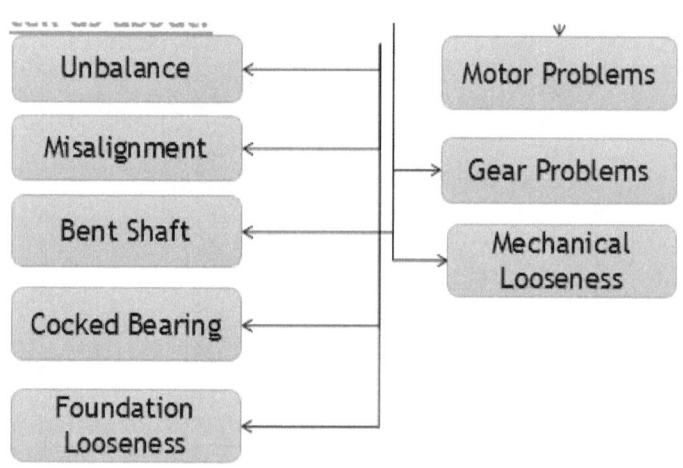

Industrial Electric Motors: By Mohammed H.A. Soliman

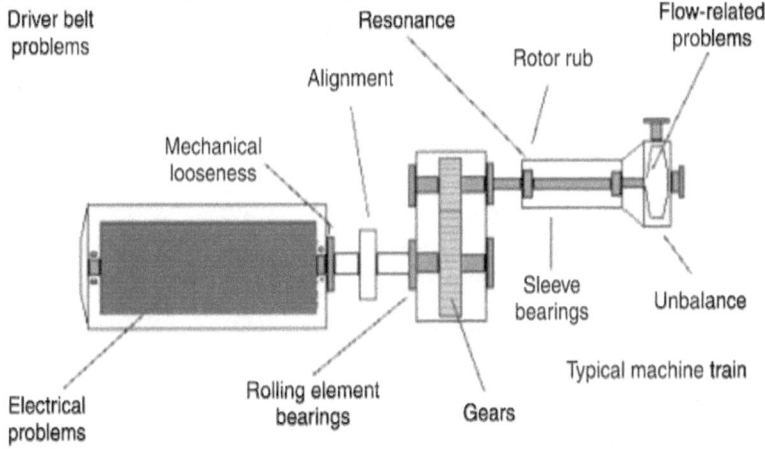

Illustrate different Machine Faults Detected by Vibration Analysis

Industrial Electric Motors: By Mohammed H.A. Soliman

Faults to be Detected by Spectrum Analysis

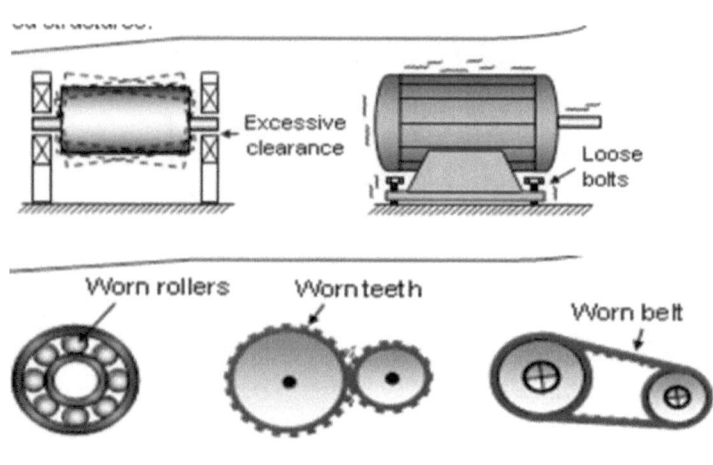

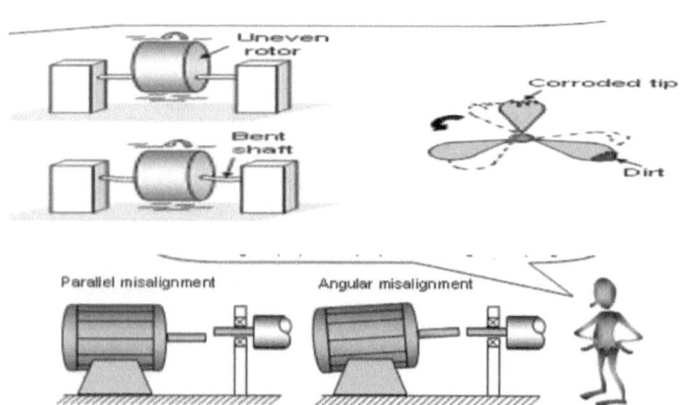

Industrial Electric Motors: By Mohammed H.A. Soliman

Predefined Spectrum Analysis Bands

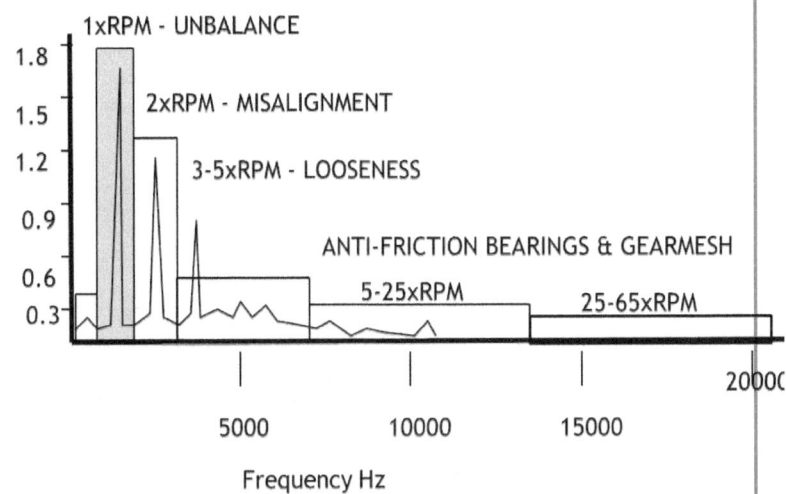

Industrial Electric Motors: By Mohammed H.A. Soliman

Construction of Spectrum Analysis

Frequency in terms of RPM	Most likely causes	Other possible causes and remarks
1x RPM	Unbalance	1) Eccentric journals, gears or pulleys 2) Misalignment or bent shaft- if high axial vibration 3) Resonance 4) Reciprocating forces 5) Electrical problems
2x RPM	Mechanical Looseness	1) Misalignment if high axial vibration 2) Reciprocating forces 3) Resonance 4) Bad belts if 2x RPM of belt
3x RPM	Misalignment	Usually a combination of misalignment and excessive axial clearances (looseness)
Less than 1x RPM	Oil whirl (less than ½ RPM)	1) Bad drive belts 2) Background vibration 3) Sub-harmonic resonance
Synchronous (A.C. Line Frequency)	Electrical Problems	Common electrical problems include broken rotor bars, eccentric rotor, unbalanced phases in poly-phase systems, unequal air gap.
2x Synch. Frequency	Torque pulses	Rare as a problem unless resonance is excited

Frequency in terms of RPM	Most likely causes	Other possible causes and remarks
High frequency (not harmonically related)	Bad anti-friction bearings	1) Bearing vibration may be unsteady- amplitude and frequency 2) Cavitations, recirculation and flow turbulence cause random, high frequency vibration. 3) Improper lubrication of journal bearings (friction excited vibration) 4) Rubbing

Frequency in terms of RPM	Most likely causes	Other possible causes and remarks
Many times RPM (harmonically related freq.)	Bad gears Aerodynamic forces Hydraulic forces Mechanical looseness Reciprocating forces	Gear teeth times RPM of bad gear Number of fan blades times RPM Number of impeller vanes times RPM May occur at 2,3,4 and sometimes higher harmonics if severe looseness

Spectrum Analysis Goals:

- Determine if a fault condition exist.
- Diagnose the fault condition.
- Determine if additional analysis is required.

- Determine severity and action required.
- Investigate root cause and provide feedback.

Industrial Electric Motors: By Mohammed H.A. Soliman

Detect electric Motor Faults by Vibration Spectrum Analysis

It's evident that defects in the bearings represent the widest source of failure to an induction motor, thus more focus was needed on bearings defects in particular.

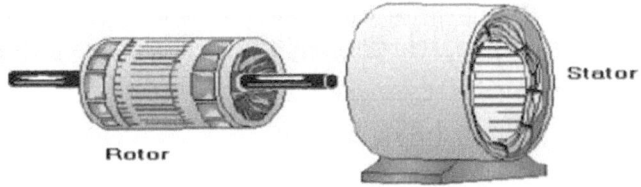

Failure	Percentage
Bearings	44%
Stator	26%
Rotor	8%
Others	22%

Industrial Electric Motors: By Mohammed H.A. Soliman

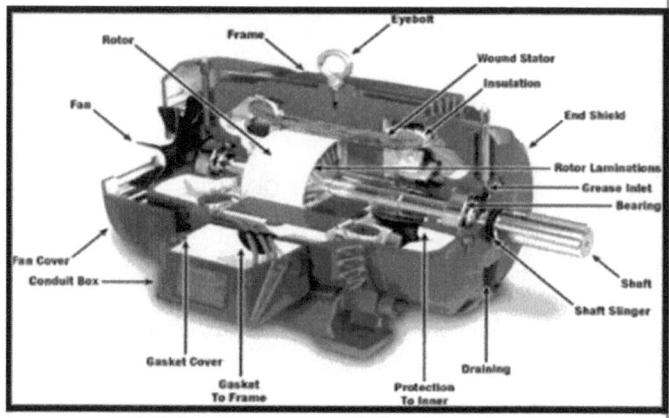

Electric motor internal components

The components of our predictive maintenance tool.

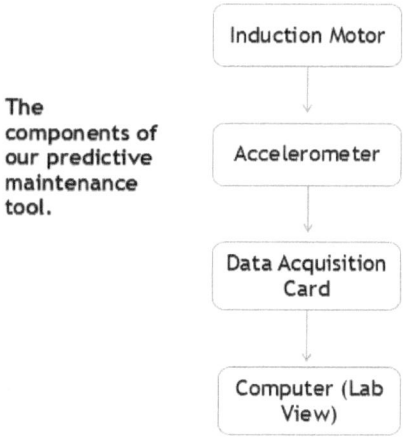

It's difficult to detect the exact reason of a bearing failure without having the

vibration test applied. The following is a list of the common defect causes:
– Ordinary wear.
– Too high ambient temperature.
– Corrosion.
– Reduced lubrication.
– Misalignment.
– Vibrations.
– Damage due to transport.
– Bearing currents from frequency converter drive.

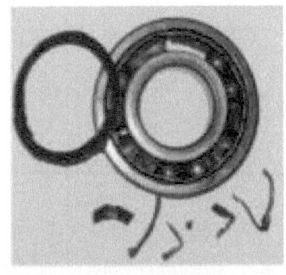

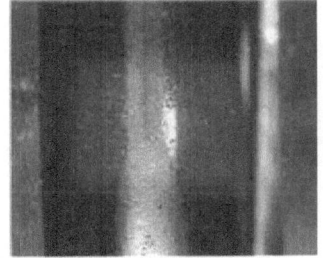

Industrial Electric Motors: By Mohammed H.A. Soliman

1. Detect the Nature of Bearing Failure

When a certain defect is present on a bearing element (example of a rough defect is shown in the above figure) an increase in the vibration levels at this frequency can be noticed, and that's why frequency-domain analysis of vibration reading is usually carried out to determine the condition of motor bearings. Frequency-domain or spectral analysis of vibration signal is the most widely used approach for bearing defect detection.

Example

1500um outer race defect

Rough defect condition in a bearing

Formula to calculate the outer race defect frequency:
BPFO (Ball Pass Frequency, Outer race):

BPFO=Nb/2(1-Bd/PDcos α) x RPM
Where:
Bd=diameter of rolling element.
PD=pitch diameter.
α=the contact angel.
Nb=no of rolling elements.
RPM= the shaft rotating frequency.

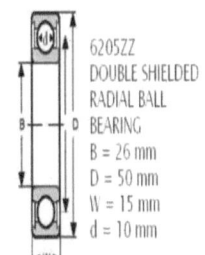

Motor bearing specifications from the tables:
Shaft rotating frequency of 24Hz
Bearing having 9 balls of diameter 8.5mm
Pitch circle diameter of 38.5mm
Contact angle a of 0.
RPM= RPM/60

Calculations = 9x24/2 X (1-0.22) =84.24Hz.

Bearing Health Condition

Industrial Electric Motors: By Mohammed H.A. Soliman

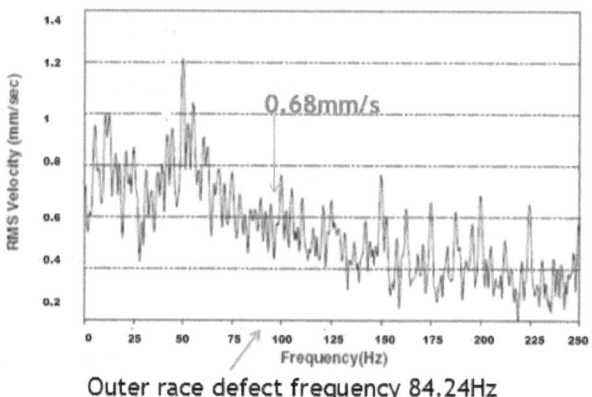

Outer race defect frequency 84.24Hz

Bearing Condition After 1 month

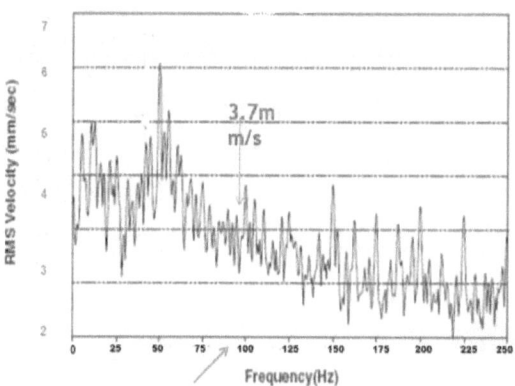

Bearing Faulty Condition

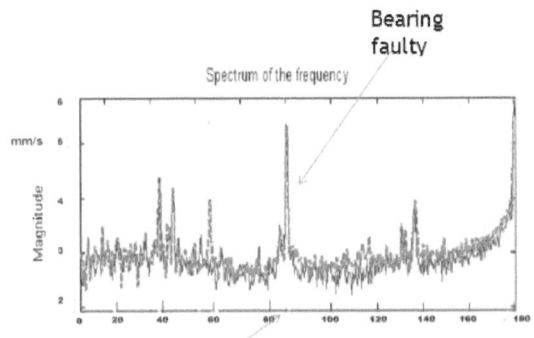

Spectrum & Vibration Analysis
BPFO defect of the bearing can be measured at 84.24hz, bearing vibration amplitude showing significant increase at the bearing outer race frequency indicating outer race defect.

Recommendation
It's highly recommended to shut down the machine to replace the defected bearing.

Industrial Electric Motors: By Mohammed H.A. Soliman

Common Bearing Failure Patterns

$$BPFI = \frac{Nb}{2}(1 + \frac{Bd}{Pd} \cos \theta) \times rpm$$

$$BPFO = \frac{Nb}{2}(1 - \frac{Bd}{Pd} \cos \theta) \times rpm$$

$$FTF = \frac{1}{2}(1 - \frac{Bd}{Pd} \cos \theta) \times rpm$$

$$BSF = \frac{Pd}{2Bd}\left[1 - \left(\frac{Bd}{Pd}\right)^2 (\cos \theta)^2\right] \times rpm$$

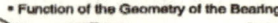

Nb = Number of Balls or Rollers
Bd = Ball / Roller diameter (inch or mm)
Pd = Bearing pitch diameter (inch or mm)
θ = Contact angle in degrees

BPFI = Ball pass frequency – Inner
BPFO = Ball pass frequency – Outer
FTF = Fundamental train frequency (Cage)
BSF = Ball spin frequency (rolling element)

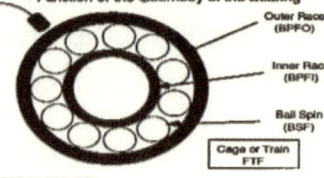

2. Detection of Different Electric Problems

The following are some terms that will be required to understand vibrations due to electrical problems:

F_L = electrical line frequency (50/60 Hz)

F_s = slip frequency = $\dfrac{2 \times F_L}{P}$ – rpm

F_p = pole pass frequency = $F_s \times P$

P = number of poles.

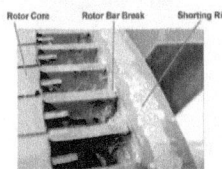

Rotor Problems
- ➤ Broken rotor bars.
- ➤ Open or shorted rotor windings.
- ➤ Bowed rotor.
- ➤ Eccentric rotor.

Industrial Electric Motors: By Mohammed H.A. Soliman

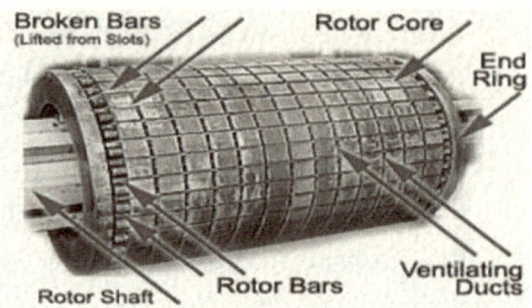

Defect rotor bars

A. Rotor Defects

Broken rotor bars.
Eccentric rotor.

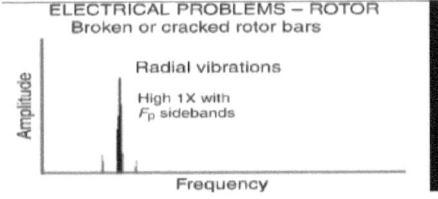

High 1X with FP sideband s

⬅ Broken rotor bars

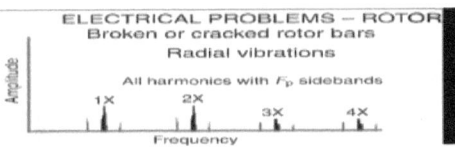

⬅ Broken rotor bars
All harmonics with FP sidebands

Example:
Motor Speed Synchronous = 1800 RPM
Motor Speed Actual = 1770 RPM
No of poles = 4
FL = 60hz
FP = 2xFL/P-RPM*P=2hz

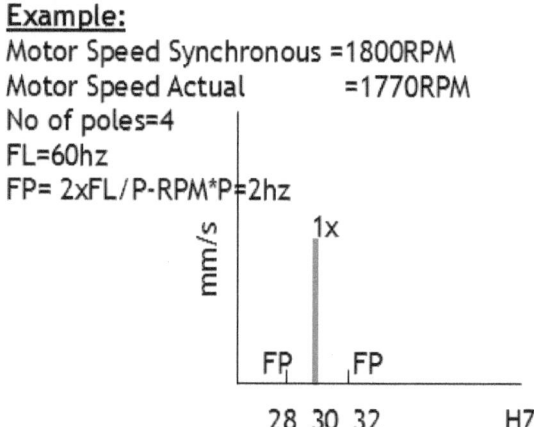

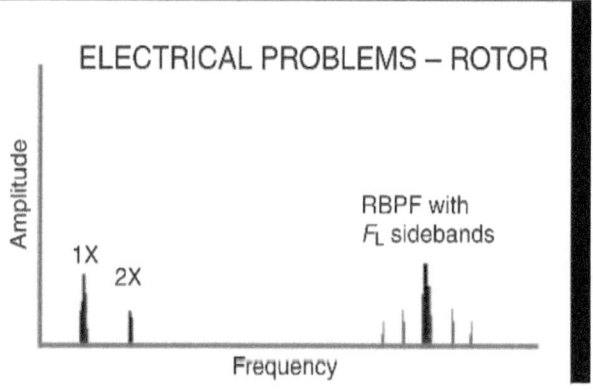

Rotor bar pass frequency, RBPF = number of rotor bars. RPM/60.

B. Eccentric Rotor

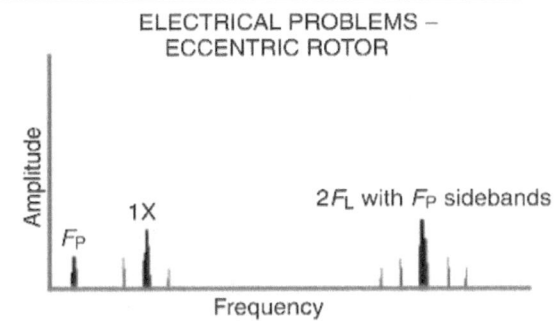

C. Stator Defects

Eccentricity
 Short lamination
 Loose iron

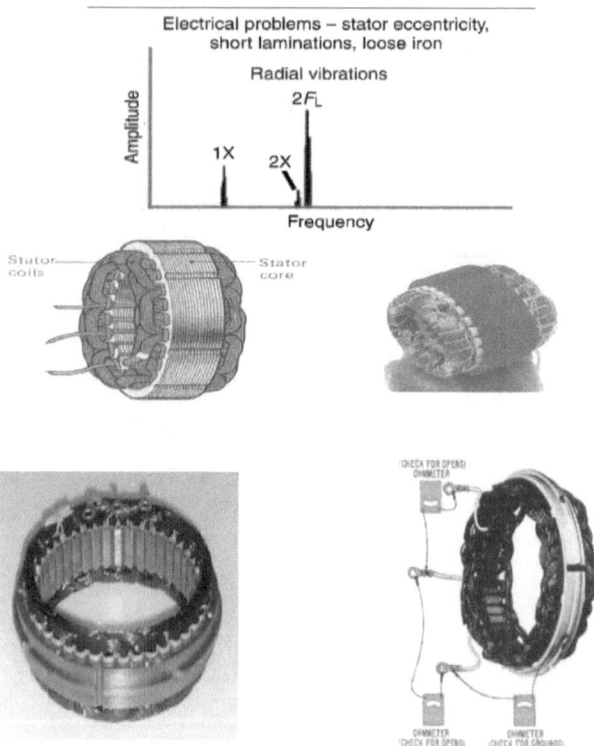

Stator Coil, Charging stator coils.

D. Phasing Problem (loose connector)

Often occurs due to broken connectors.

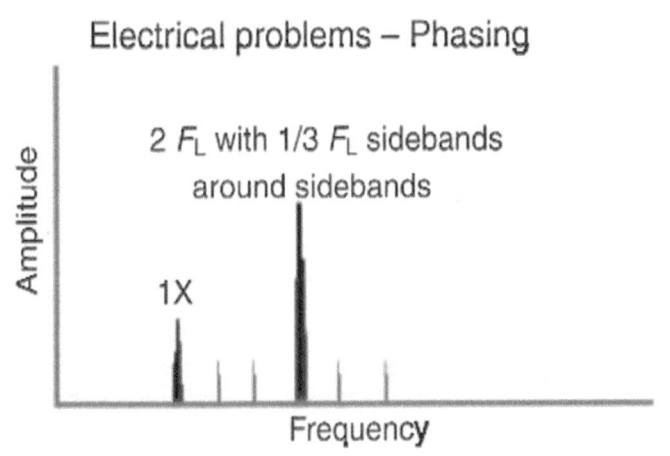

E. Synchronous Motors (Loose stator coils)

CPF = number of stator coils. RPM/60

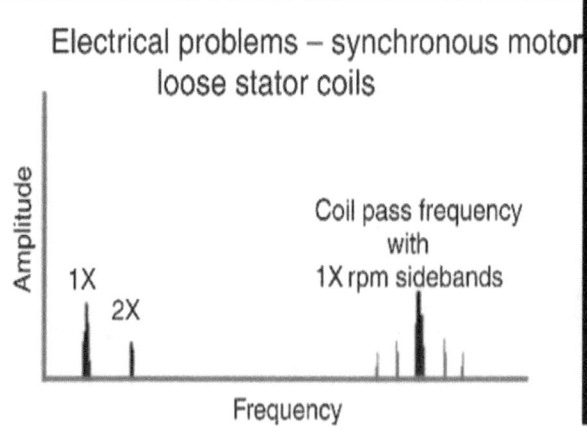

F. DC motor problems

Possible reasons:
- Broken field windings.
- Bad SCRs.

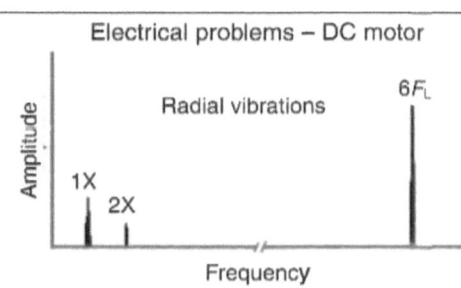

- Loose connection.
- Loose or blown fuses.
- Shorted control cards.

3. Using Vibration Spectrum Analysis to Detect Machine Looseness

If we consider any rotating machine, mechanical looseness can occur at three locations:
1. Internal assembly looseness.
2. Looseness at machine to base plate interface.
3. Structure looseness.

I. Internal Assembly Looseness

This category of looseness could be between a bearing liner in its cap, a sleeve or rolling element bearing, or an impeller on a shaft. It is normally caused by an improper fit between component parts, which will produce many harmonics in the FFT due to the nonlinear response of the loose parts to the exciting forces from the rotor.

Industrial Electric Motors: By Mohammed H.A. Soliman

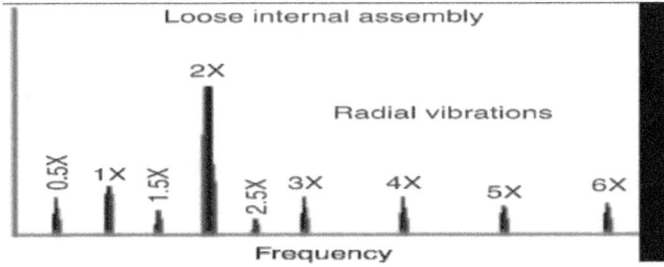

Industrial Electric Motors: By Mohammed H.A. Soliman

II. Looseness between Machine to Base Plate

This problem is associated with loose pillow-block bolts, cracks in the frame structure or the bearing pedestal.

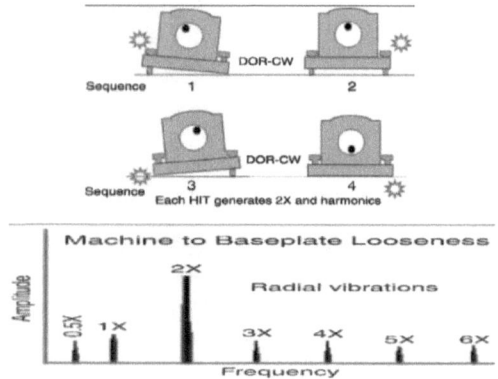

III. Structure Looseness

This type of looseness is caused by structural looseness or weaknesses in the machine's feet, baseplate or foundation.

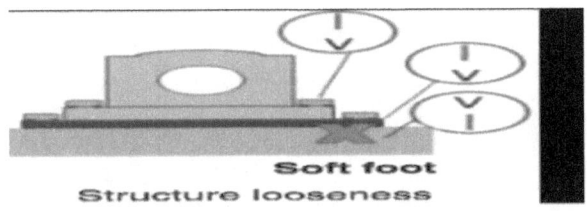
Soft foot
Structure looseness

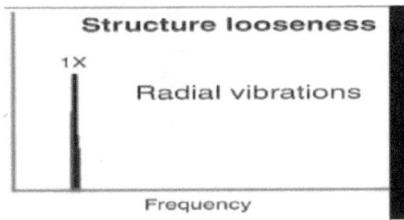

When the soft foot condition is suspected, an easy test to confirm for it is to loosen each bolt, one at a time, and see if this brings about significant changes in the vibration. In this case, it might be necessary to re-machine the base or install shims to eliminate the distortion when the mounting bolts are tightened again.

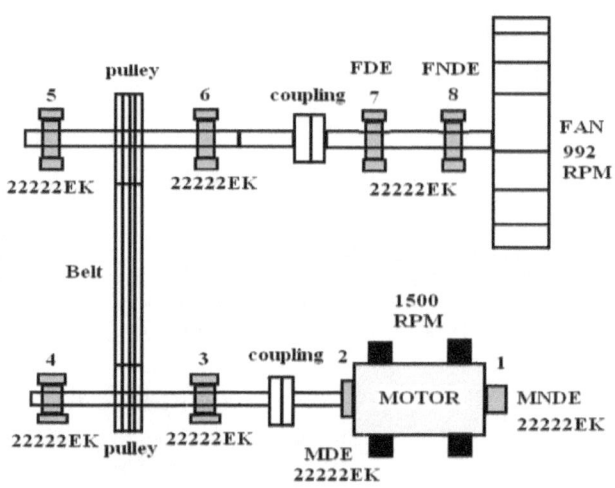

de-gasing fan15B802

Vibration Reading:

Bearing 3-H	= 5.965
Bearing 3-V	= 8.632
Bearing 4-H	= 16.042
Bearing 4-V	= 12.828

Industrial Electric Motors: By Mohammed H.A. Soliman

Bearing 5-H   = 5.83
Bearing 5-V   = 5.914
Bearing 6-H   = 7.812
Bearing 6-V   = 6.505
FDE-H   = 6.512
FDE-V   = 11.878
FNDE-H   = 4.805
FNDE-V   = 17.869

Spectrum for bearings 4 & 5

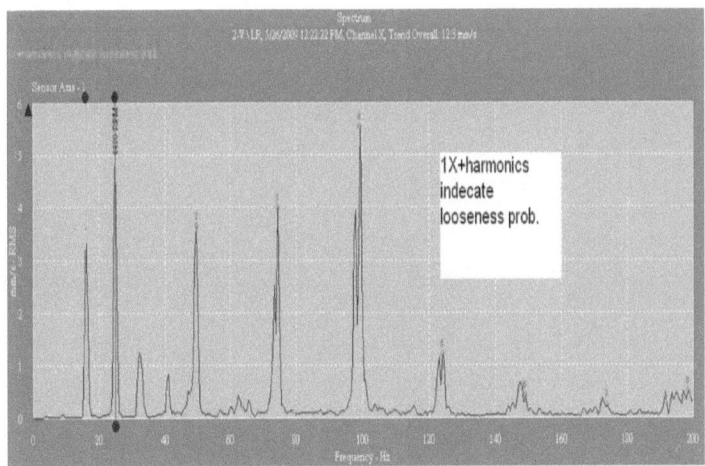

Spectrum Analysis:
Check Bearing no. (4) for fixation problem with the base. (Retighten bearing housing bolts & check bearings internal clearance).

Check Bearing no. (5) For fixation problem with the base. (Retighten bearing housing bolts & check bearings internal clearance).

Maintenance work to be performed:
1-Retightining of bearing bolts & housing.
2-Monitor bearing condition with some other methods (use ultrasonic).

Literature References

American Vibration Institute.

MOBIUS Institute (ilearn).

Scheffer, C. and Girdhar, P. 2004. Practical Machinery Vibration Analysis and Predictive Maintenance: Newnes; 1st Edition Predictive.

Soliman, M.H.A. 2020. Machine Reliability and Condition Monitoring: A Comprehensive Guide to Predictive Maintenance. Personal-lean.org

Soliman, M.H.A. 2020. Industrial Applications of Infrared Thermography. Lulu Press and KDP.

Soliman, M.H.A. 2020. Ultrasound Analysis for Condition Monitoring. Personal-lean.org

Soliman, M.H.A. 2020. A Practical Guide to FMEA. Personal-lean.org

Soliman, M.H.A. 2014. Analyzing Failure to Prevent Problems. Industrial Management.

Soliman, M.H.A. 2020. Vibration Basics and Machine Reliability Simplified: A Practical Guide to Vibration Analysis. Personal-lean.org.

Soliman, M.H.A. 2020. Brainstorming for Problems Solving: How Leaders Can Achieve a Successful Brainstorming Session. Personal-lean.org

Soliman, M.H.A. 2021. Risk Assessment Using FMEA: A Case of Reliable Improvement. Personal-lean.org.

Soliman, M.H.A. 2020. Machinery Oil Analysis & Condition Monitoring: A Practical Guide to Sampling and Analyzing Oil to Improve Equipment Reliability. Personal-lean.org.

Don't miss out!

Visit the website below and you can sign up to receive emails whenever Mohammed Hamed Ahmed Soliman publishes a new book. There's no charge and no obligation.

https://books2read.com/r/B-A-VCQM-ZKMQD

BOOKS 2 READ

Connecting independent readers to independent writers.

Did you love *Industrial Electric Motors: Installation, Running, Advanced Maintenance and Reliability*? Then you should read *Industrial Power Transformers: Selection, Installation, Advanced Maintenance and Reliability* by Mohammed Hamed Ahmed Soliman!

This book aims to explain the best ways to do work that is usually done to avoid issues with transformers. This book covers everything about choosing and storing transformers. It also talks about advanced methods for checking transformers using predictive maintenance or condition monitoring. It also includes a real example of using FMEA to make power transformers more reliable in a system or production process. The techniques in this book are not for making big changes to repair a transformer. However, many things are done as part of regular procedures. Maintenance and big transformer repair could be the same. We can do the tasks to take care of the transformer if it's not too broken. The advice in this book is similar to the suggestions that companies give for their products. If you need to know how to do something, the person in charge should check the instruction book from the company that made the product. Regularly check and fix small problems to keep transformers in good condition. Also follow special care instructions. Also, if the machine is set up and used the right way, it will keep working for a long time without any issues.

Read more at https://www.personal-lean.org/.

Also by Mohammed Hamed Ahmed Soliman

Toyota Production System Concepts
The Toyota Way to Effective Strategy Deployment Using Hoshin Kanri
Heijunka and the Art of Production Leveling
Jidoka - Automation with Human Intellegince
Lean Pull System and Kanban
Understanding OEE in Lean Production
How to Use PDCA Cycle of Improvement to Develop Lean Leaders
SMED – How to Do a Quick Changeover?
How Lean Can improve Healthcare? A Brief Guide on Eliminating Waste and Identifying Areas for Improvement
Identifying Mura-Muri-Muda in the Manufacturing Stream
Takt Time - Understanding the Core Principle of Lean Manufacturing
Understanding Lean Financial Accounting
What Andon Truely is in Lean Manufacturing?
Standardized Work is a Goal - Not Just a Tool in Lean Practices

How to Develop Mission, Strategy, Goals and Values That Fit with Company's Vision Statement
Lean Approach to Cost-Benefit Analysis
Genchi Genbutsu Process – The Role of Gemba in Lean Management and Value Creation
Toyota's Approach to Developing and Coaching Leaders
How to Create Continuous Production Flow?
Lean Culture - How Toyota Encourages Employees to Embrace Lean Behaviors and Practices
5S- The True Mean to Enhance Productivity and Work Value for Customers
Understanding the Toyota Production System's Genetics
The Guidebook to Toyota's 13 Pillars System - Series Books 7 to 17
The Guidebook to Toyota's Corporate Strategy and Leadership – Series Books 1 to 6
What are The Improvement Kata and Coaching Kata?
Process Mapping the Toyota Way
Application of Lean in Non-manufacturing Environments - Series Books 18 to 19

Standalone
Hoshin Kanri: How Toyota Creates a Culture of Continuous Improvement to Achieve Lean Goals
The Seven Deadly Wastes and How to Remove Them from Your Business: The Heart of the Toyota Production System
Overall Equipment Effectiveness Simplified: Analyzing OEE to find the Improvement Opportunities

Machinery Oil Analysis & Condition Monitoring : A Practical Guide to Sampling and Analyzing Oil to Improve Equipment Reliability

Practical Guide to FMEA : A Proactive Approach to Failure Analysis

Industrial Applications of Infrared Thermography: How Infrared Analysis Can be Used to Improve Equipment Inspection

Ultrasound Analysis for Condition Monitoring: Applications of Ultrasound Detection for Various Industrial Equipment

Brainstorming for Problems Solving: How Leaders Can Achieve a Successful Brainstorming Session

Vibration Basics and Machine Reliability Simplified : A Practical Guide to Vibration Analysis

Gemba Walks the Toyota Way : The Place to Teach and Learn Management

Jidoka: The Toyota Principle of Building Quality into the Process

Turning PDCA into a Routine for Learning

Toyota Healthcare: 7+1 Types Of Waste

Kanban the Toyota Way: An Inventory Buffering System to Eliminate Inventory

Takt Time: A Guide to the Very Basic Lean Calculation

5S: A Practical Guide to Visualizing and Organizing Workplaces to Improve Productivity

Machine Reliability and Condition Monitoring: A Comprehensive Guide to Predictive Maintenance Planning

The Ultimate Guide to Successful Lean Transformation: Top Reasons Why Companies Fail to Achieve and Sustain Excellence through Lean Improvement

Toyota Standard Work: The Foundation of Kaizen

Risk Assessment Using FMEA: A Case of Reliable Improvement

5S: A Practical Guide to Visualizing and Organizing Workplaces to Improve Productivity

Hoshin Kanri: How Toyota Creates a Culture of Continuous Improvement to Achieve Lean Goals

Industrial Applications of Infrared Thermography: How Infrared Analysis Can be Used to Improve Equipment Inspection

Ultrasound Analysis for Condition Monitoring: Applications of Ultrasound Detection for Various Industrial Equipment

Manufacturing Wastes Stream: Toyota Production System Lean Principles and Values

The Ultimate Guide to Successful Lean Transformation: Top Reasons Why Companies Fail to Achieve and Sustain Excellence through Lean Improvement

The Problem Solving Kata as a Tool for Culture Change: Building True Lean Organizations

Lean Healthcare: Enhancing the Patient Care Process while Eliminating Waste and Lowering Costs

Heijunka: The Leveling Art of the Japanese Auto Industry

Creating a One-Piece Flow and Production Cell: Just-in-time Production with Toyota's Single Piece Flow

A Complete Guide to Just-in-Time Production: Inside Toyota's Mind

Poka Yoke - Design for Great Quality

Design Review Based on Failure Mode: How Toyota Engineers Use an Advanced Model of FMEA to Improve Product Reliability (Toyota's GD3)

Industrial Power Transformers: Selection, Installation, Advanced Maintenance and Reliability

Industrial Electric Motors: Installation, Running, Advanced Maintenance and Reliability

Watch for more at https://www.personal-lean.org/.

About the Author

Mohammed Hamed Ahmed Soliman is an industrial engineer, consultant, university lecturer, operational excellence leader, and author. He works as a lecturer at the American University in Cairo and as a consultant for several international industrial organizations.

Soliman earned a Bachelor's of science in Engineering and a Master's degree in Quality Management. He earned post-graduate degrees in Industrial Engineering and Engineering Management. He holds numerous certificates in management, industry, quality, and cost engineering.

For most of his career, Soliman worked as a regular employee for various industrial sectors. This included crystal-glass making, fertilizers, and chemicals. He did this while educating people about the culture of continuous improvement. Soliman has more than 15 years of experience and proven track record of achieving high levels of operational excellence to a broad range of business operations including manufacturing, service and healthcare. He has led several improvement projects within leading organizations and defined a lot of savings in the manufacturing wastes stream.

Soliman has lectured at Princess Noura University and trained the maintenance team in Vale Oman Pelletizing Company. He has been lecturing at The American University in Cairo for 8 years and has designed and delivered 40 leadership and technical skills enhancement training modules. In the past 4 years, Soliman's lectures have been popular and attracted a large audience of over 200,000 people according to SlideShare's analysis.. His research is one of the most downloaded works on the Social Science Research Network, which is run by ELSEVIER. His research is one of the most downloaded works on the Social Science Research Network, which is run by ELSEVIER.

Soliman is a senior member at the Institute of Industrial and Systems Engineers and a member with the Society for Engineering and Management Systems. He has published more than 60 publications including articles in peer reviewed academic journals and international magazines. His writings on lean manufacturing, leadership, productivity, and business appear in Industrial Engineers, Lean Thinking, Industrial Management, and Sage Publications. Soliman's blog is www.personal-lean.org.

Read more at https://www.personal-lean.org/.

About the Publisher

Personal-lean is dedicated to publish high quality educational content, assessment, training in the filed of business for various industrial sectors. And is a growing educational organization, with products and services in various countries.

www.ingramcontent.com/pod-product-compliance
Lightning Source LLC
Chambersburg PA
CBHW020633220526
45464CB00001B/137